PASOLD STUDIES IN TEXTILE HISTORY, 9

Fashion's Favourite

FASHION'S FAVOURITE: THE COTTON TRADE AND THE CONSUMER IN BRITAIN, 1660–1800

BEVERLY LEMIRE

PASOLD RESEARCH FUND
OXFORD UNIVERSITY PRESS
1991

Oxford University Press, Walton Street, Oxford OX2 6DP
Oxford New York Toronto
Delhi Bombay Calcutta Madras Karachi
Petaling Jaya Singapore Hong Kong Tokyo
Nairobi Dar es Salaam Cape Town
Melbourne Auckland
and associated companies in
Berlin Ibadan

Oxford is a trade mark of Oxford University Press

Published in the United States
by Oxford University Press, New York

© The Pasold Research Fund 1991

All rights reserved. No part of this publication may be reproduced,
stored in a retrieval system, or transmitted, in any form or by any means,
electronic, mechanical, photocopying, recording, or otherwise, without
the prior permission of Oxford University Press

British Library Cataloguing in Publication Data
Data available

Library of Congress Cataloging in Publication Data
Lemire, Beverly, 1950–
Fashion's favourite: the cotton trade and the consumer in
Britain, 1660–1800/Beverly Lemire.
1. Cotton textile industry—Great Britain—History. 2. Cotton
fabrics—Great Britain—History. 3. Cotton trade—Great Britain—
History. 4. Clothing trade—Great Britain—History. I. Title.
HD9881.5.L46 1992
338.4'767721'0941—dc20   91-26790
ISBN 0-19-921062-4

Typeset by Cambrian Typesetters, Frimley, Surrey
Printed and bound in
Great Britain by Biddles Ltd,
Guildford and Kings Lynn

To my family with love

ACKNOWLEDGEMENTS

The links between Britain's cotton trade, popular fashion, and the consumer in the era leading up to industrialization have preoccupied me for many years. I first wondered about these connections while studying for an exam as an undergraduate—a fruitful diversion that resulted in many more years of a different sort of study. In the course of this work I incurred debts, both academic and personal, that I am happy to acknowledge here. My first and greatest debt is to my husband, Morris, and daughter, Shannon. I have been inspired, encouraged, and enriched by their support over many years. Their sincere empathy and adventurous spirit sustained me.

Assistance and support of an academic and personal nature were extended to me many times and in many ways over the course of my research into this topic. I first began working on this subject at the Department of History of the University of Guelph. There I received unwavering encouragement to pursue my interests, despite the paucity of materials at that time on the history of consumer practice. In particular I extend my thanks to my former supervisor, David Murray. The financial support of the Canadian Association of University Teachers and the Ontario Government measurably assisted my research.

The Commonwealth Association of Universities provided invaluable financial support during the period of my doctoral research in Oxford. I would like to acknowledge also the many facilities made available to me by Balliol College. John Prest, in particular, gave unstintingly of his time and encouragement over many years. My debt to Stanley Chapman is considerable, not only for his efforts as my doctoral supervisor, but also for his reliable and penetrating advice during the added years of research that went into the preparation of this book. He offered generous suggestions and key insights over the course of this project. Peter Mathias also counselled and advised as I launched my research in broader channels. I thank him for his time and attention. Madeleine Ginsburg, formerly of the Department of Textiles of the Victoria and Albert Museum, generously shared her expertise in material evidence.

Former colleagues from the University of Lethbridge, and present colleagues at the University of New Brunswick, welcomed me into a rich and stimulating academic environment—an incalculable asset. Malcolm Greenshields, Brent Shaw, and Jim Tagg from the former institution, and Gillian Thompson, Gail Campbell, Gary Waite, Steve Turner, Bill Acheson, Phil Buckner, Ernie Forbes, Marc Milner, David Frank, and Stephen Patterson offered advice and analyses of numerous historical questions that enriched my understanding of history and the pursuit of historical research.

To Joan Thirsk I also owe thanks. I benefited from her valuable criticisms, insights, and encouragement in the development of this volume. This timely assistance was graciously and generously provided over the last several years. Negley Harte has also provided a continuing influence on the progress of this book through several drafts, and as the representative of the Pasold Research Fund has provided contributions for which I am grateful.

Segments of chapters 2 and 5 were published in modified form in *Textile History* and the *Journal of British Studies* respectively, and are presented here in a considerably revised format.

Librarians and archivists helped at every level of my research. I would like to thank especially the librarians of the Bodleian Library, and in particular Julie Anne Wilson of the John Johnson Collection. Thanks are extended also to the archivists of the Guildhall Library, the Kent County Record Office, and the Manchester Public Library, as well as the many other archivists on whom I relied. In the face of continuing funding shortfalls, librarians and archivists display an impressive resourcefulness.

In the long gestation-period inherent in a work such as this the tangible and intangible sustenance provided by friends and family cannot be overestimated. My mother and father rank first among those to whom I owe thanks for continuing and unwavering support. Similarly, I thank my sister Catherine for the suggestions and new approaches she proposed over the years, and my brother Martin for his timely support. In addition, I have benefited for many years from the intellectual energy generated by Jane Stapleton and Lynne Hancock. I recognize that my work was advanced through their continuing stimulation and unremitting support. My thanks are proferred also to Linora Lawrence. The collective backing of family, friends, and colleagues, both those mentioned here and those left unnamed, contributes immeasurably to the momentum of research, without which my personal and academic life would be diminished.

CONTENTS

LIST OF ILLUSTRATIONS x

LIST OF TABLES xii

    Introduction 1

1. Popular Fashion and East India Trade: The Advent of Cotton Textiles 3

2. Domestic Demand in the Eighteenth Century: The British Market for Cotton Textiles 43

3. 'Who Will Buy?' The Consumer Process and the Cotton Industry 77

4. Distribution and Sale of British Cottons in the Home Market 115

5. Marketing Fashion: Cotton Clothing and the Ready-made Clothes Trade 161

APPENDICES
1. Lancashire Textiles, 1775–1785 201
2. Ownership of Clothing from Contemporary Court Records 203
3. Travels of Joseph Harper 227
4. Customers Served by the Anonymous Manchester Firm, 1773–1779 229

INDEX 233

LIST OF ILLUSTRATIONS

(between pages 84–85)

1. 'A Particular of Goods'. (Bodleian Library: John Johnson Collection, East India Company, Box 1)
2. Calico printer. (Bodleian Library: John Johnson Collection, Trades and Professions, Box 3)
3. 'The Case of the Weavers'. (Bodleian Library: John Johnson Collection, Commerce, Box 4)
4. Printed Indian gown, c.1740. (Textile Department, Royal Ontario Museum, Toronto)
5. Printed Indian gown, c.1740. (Textile Department, Royal Ontario Museum, Toronto)
6. Trade-cards: Cotes's Manchester Warehouse. (Bodleian Library: John Johnson Collection, Trade Cards, Box 22)
7. Romanis, Hosier and Manufacturer. (Bodleian Library: John Johnson Collection, Mens Clothing, Box 3)
8. Hannah Tatum. (Bodleian Library: John Johnson Collection, Trade Cards, Box 6)
9. Price, and Co. (Bodleian Library: John Johnson Collection, Mens Clothing, Box 3)
10. Brown, Taylor and Habit-maker. (Bodleian Library: John Johnson Collection, Trade Cards, Box 27)
11. Street-trader of second-hand clothes. (Guildhall Library, Corporation of London)
12. Jew purchasing old clothes. (Guildhall Library, Corporation of London)
13. 'The Mop Trundler'. (Bodleian Library: John Johnson Collection, Trades and Professions, Box 4)
14. 'Quite Ripe Sir'. (Bodleian Library: John Johnson Collection, Trades and Professions, Box 4)

(between pages 180–181)

15. Child's embroidered muslin dress, late eighteenth century. (Textile Department, Royal Ontario Museum, Toronto)
16. Child's embroidered muslin dress, late eighteenth century. (Textile Department, Royal Ontario Museum, Toronto)

List of Illustrations

17. Page from the Barbara Johnson Sample Book. (Courtesy of the Board of Trustees of the Victoria and Albert Museum)
18. Trade-card, Greenfield's Warehouse, London. (Guildhall Library, Corporation of London)
19. Trade-card, Gedge, Linen-draper, London. (Guildhall Library, Corporation of London)
20. 'Charlotte Bateman'. (Bodleian Library: *Lady's Magazine*, Per. 2705.e.1279)
21. Six engravings of dress for the year, from almanacks. (Bodleian Library: John Johnson Collection, Trade Cards, Box 8)
22. 'Bold Stroke for a Husband'. (Bodleian Library: *Lady's Magazine*, Per. 2705.e.1279)
23. Trade-card, Gown Warehouse, Temple Exchange. (Bodleian Library: John Johnson Collection, Trade Cards, Box 13)
24. Hogarth trade-card. (Bodleian Library: John Johnson Collection, Trade Cards, Box 9)
25. Trade-card, John Keet, Portsmouth. (Bodleian Library: John Johnson Collection, Trade Cards, Box 21)
26. Allin's Cheap Clothes Warehouse, Birmingham. (Bodleian Library: John Johnson Collection, Mens Clothing, Box 1)

LIST OF TABLES

2.1.	Estimates of population for England, 1701–1801	44
2.2.	Age of first marriage through the eighteenth century	45
2.3.	Estimates of nominal earnings for eighteen occupations, 1755–1851: adult males, England and Wales	49
2.4.	Retained imports of raw cotton	54
3.1.	Lancashire Cotton textiles in Holker's report, $c.$1750	80
4.1.	London tradesmen operating as wholesale distributors	131

MAP

Location of customers supplied by the unknown Manchester firm 1773–79   127

INTRODUCTION

THE history of the cotton industry in eighteenth-century Britain captured the attention of historians over the course of this and the last century. Much of this history has been well documented. The stories of the individual inventors, their trials and successes, have been recounted at some length in the earliest hagiography.[1] Later studies focused on the complex and unprecedented sequence of events arising from the first mechanized trade in what has come to be called the Industrial Revolution. The British cotton industry displayed a capacity of production unmatched up to that time. The rising capacity of the mills and the product itself fascinated historians, as did the distinctive history of British trade in cottons throughout the world. The focus of most investigations of the cotton industry has centred on technological and industrial developments, business histories, and the markets abroad for cotton products.[2] Readers of this book are assumed to be well aware of the stages of development of the cotton industry: starting with the spread of cottage production before the turn of the century, to the rise of the spinning mills from the 1770s onwards, leading to the development of industrial manufacturing areas in the midlands and north-west, and the rise of cottonopolis itself—Manchester. A history so well known need not be chronicled again here.[3]

The purpose of this volume is to address one of the unconsidered elements of the cotton trade. Although so much is known of the workings of factories and family enterprise, little time has been spent assessing the relationship between those who bought cotton textiles

---

[1] Edward Baines, *History of the Cotton manufacture in Great Britain* (1835), is one of the best examples of these early works describing the growth of the cotton industry in very uncritical terms.
[2] In addition to the studies cited in Chap. 1, n. 7 below, see: D. A. Farnie, *The English Cotton Industry and the World Market* (1979); Mary B. Rose, *The Gregs of Quarry Bank Mill* (1986); R. S. Fitton, *The Arkwrights: Spinners of Fortune* (1989).
[3] The most succinct survey of the cotton industry, along with a survey of the literature, can be found in S. D. Chapman, *The Cotton Industry in the Industrial Revolution*, 2nd edn. (1987).

and the products, both Indian and British, that were supplied to them. This history begins with the arrival of East India calicoes on British shores in the seventeenth century and concludes with the preeminence of cotton clothing throughout the population at the end of the eighteenth century. Recent accounts of the manner in which the market was served have revealed the strength of demand in Britain and the importance to most ranks of the niceties of life.[4] The British market was a cardinal factor in the development of the cotton industry. It was the largest area of free trade in Europe and, as the eighteenth century approached, had a level of prosperity that was marked by only the occasioned crisis. Yet the characteristics of this market and its relationship with the cotton manufacturers has been too little studied. The choices made by British consumers directed the way in which the cotton industry would grow and determined the sorts of goods that would be produced. The products of this industry and the manner in which they were sold and used marked new patterns of consumer activity that altered the appearance of the nation. These new cotton commodities came to symbolize the later industrial age.

[4] Joan Thirsk, *Economic Policy and Projects* (1979); Margaret Spufford, *The Great Reclothing of Rural England* (1984); Neil McKendrick et al., *The Birth of a Consumer Society* (1982); Lorna Weatherill, *Consumer Behavior and Material Culture in Britain, 1660–1760* (1988).

1
POPULAR FASHION AND EAST INDIA TRADE: THE ADVENT OF COTTON TEXTILES

INTRODUCTION

NEW consumer forces were at work in England in the late seventeenth century; these forces were manifested in the craze for calicoes and other sorts of cottons wrought in the East Indies. The arrival of cotton textiles in the English market brought to a comparatively prosperous population a new sort of commodity, another of the lighter fabrics so much in demand in the early industrial period. Since the Middle Ages, the progressive spread of cotton textiles in Europe was linked to the expanding public appetite for inexpensive but fashionable dress and furnishings.[1] The predilection for newfangled wares was well documented in the sixteenth century. The appearance of ever-more shiploads of cheap, attractive merchandise during the seventeenth century reflected the new wants among a broad spectrum of the population. 'He that considers how wonderfully Fashions prevail on this Nation may soon satisfie himself how things of little value come to be prized, and to justle out those of greater worth', bemoaned one writer at the end of the seventeenth century. Reproaches were directed not only at the 'gallants' but also at the many poorer sorts, lower in rank and limited in income, 'yet . . . willing to imitate the Rich'.[2] In spite of opposition, the urge to follow fashion grew among the middling and lower ranks, as too did the ambition among British artisans to replicate and manufacture cheap copies of novel textiles. The rise of the British cotton industry in the eighteenth century began

---

[1] Maureen Fennell Mazzaoui noted that: 'Within the total spectrum of medieval textile production, cotton manufacture stands out as the only major export industry geared to the output of low-priced goods for popular consumption with profits heavily dependent upon volume of turnover.' *The Italian Cotton Industry in the Later Middle Ages, 1100–1600* (1981), 60.

[2] John Cary, *A Discourse concerning the East-India Trade, shewing it to be Unprofitable to the Kingdom of England . . .*, n.d., 4.

with the arrival on British shores of bales of oriental textiles. There was, however, no smooth development of either the trade or the industry. Some of the innovative commercial enterprises found little support among traditional élites, who hoped to restrict or prohibit those aspects that seemed to threaten established concerns. The startling success of the East Indian fabrics took the wool merchants by surprise, challenging conventional expectations of trade, and seemingly jeopardized the social hierarchy with levelling fashions. Calicoes, chintzes, muslins, and silks were damned as 'inconsistent with the Prosperity of our England Manufactures'.[3] Competing interests converged around the issue of calico sales in England, and the apparent resolution of the contest produced unforeseen results.

An inflexible mercantile theory convinced legislators and commentators alike of the undesirability of excessive popular consumption of novelties, while the apparent danger to public standards convinced many that for both moral and economic reasons these vanities, luxuries, and novelties should be suppressed.[4] Notwithstanding the recurrent outcries from moralists and mercantilists of various hues, the East India traders could not be hedged-in, neither could the public cravings be extinguished. The older mercantilist and moral doctrines were undercut. The clamour against the perceived extravagance continued throughout the later seventeenth century as the habits of various ranks diverged further from the ideal.

Cotton textiles appeared to arrive suddenly and to acquire a phenomenal popularity almost overnight, overturning many of the cost-restraints that previously had limited popular fashions. This assessment was substantially accurate. The unexpected appearance of a precursor to Indian textiles, the New Draperies, led D. C. Coleman to suggest that 'the precise nature of a "new product" seems to need careful consideration', adding that 'economic causes alone may be inadequate to explain either the acceptance or the diffusion of the new product'.[5] Evolving patterns of consumption similarly remain largely

---

[3] J. Asgill, *A Brief Answer to a Brief State of the Question, Between the Printed and Painted Callicoes, and the Woollen and Silk Manufactures* . . . (1720), 3–4.

[4] Joyce Appleby, 'Ideology and Theory: The Tension between Political and Economic Liberalism in Seventeenth-Century England', *American Historical Review*, 81 (1976), 499–515, suggests the way in which theories of economic liberalism challenged the political order of late seventeenth-century England, being perceived as a threat to the existing political structure.

[5] D. C. Coleman, 'An Innovation and its Diffusion: The "New Draperies"', *Economic History Review*, 2nd series, 22 (1969), 429.

unmapped as they developed within pre-industrial and early industrial society. Neil McKendrick has pointed out the absence of specialized studies of consumer industries, noting that in fashion, and its relations to consumer demand, 'the interlocking relationship of these different developments (and the new explanations of the rise in home demand) have rarely been studied as a whole'.[6] Many aspects of the cotton trade have been examined by historians.[7] The most striking omission has been the absence of any detailed investigation of the growth in demand for cottons. Cotton textiles were the choice of millions of people over generations, in varying circumstances. The role played by popular fashions in the development of this consumer trade requires close attention. This book attempts to chart the course of cottons in Britain and the influence of fashion in the developing prominence of cotton textiles in the national economy.

GENESIS OF POPULAR FASHION IN DRESS

Fashions might be set at court, but in late sixteenth-century England they spread in the streets. A novelty, worn on the streets of London, was like a pebble thrown into a pool, the effects rippling inexorably through society after the initial splash. In 1564, a pair of fine worsted hose worn in the London streets by a merchant from Mantua inspired an astute apprentice to ask the owner of the stockings if the hose might be borrowed in order to have them copied. The facsimile was reputed to be the first pair of worsted stockings made in England. 'Within a few years', Stow recounts, 'began the plenteous making both of jersey and woollen stockings, so in a short space they waxed common'.[8] So great was the desire of the English for more and newer products that a new stocking industry thrived from its inception in the sixteenth century.

---

[6] Neil McKendrick *et al.*, *The Birth of a Consumer Society* (1982), 42, note.

[7] Some of the earlier historians include George M. Daniels, *The Early English Cotton Industry* (1920); George Unwin, *Samuel Oldknow and the Arkwrights* (1924); the classic study by A. P. Wadsworth and Julia de Lacy Mann, *The Cotton Trade and Industrial Lancashire* (1931); Arthur Redford, *Manchester Merchants and Foreign Trade* (1934); while among more recent works are S. D. Chapman, *The Early Factory Masters* (1967); M. M. Edwards, *The Growth of the British Cotton Trade, 1780–1815* (1967); and S. D. Chapman and S. Chassagne, *European Textile Printers in the Eighteenth Century* (1981).

[8] J. Stow, *The Annales or Generall Chronicle of England* (1615), 869, quoted in Joan Thirsk, 'The Fantastical Folly of Fashion: The English Stocking Knitting Industry, 1500–1700', in N. B. Harte and K. G. Ponting (eds.), *Textile History and Economic History* (1973), 55.

The prosperity of this industry was matched by others. New products were a prominent feature of the period, with concomitant disruptions of traditional manufacturing sectors in England. The New Draperies are the most widely known example of a product that flourished as a result of altered tastes, resulting in the decline of the old woollen manufacture. The proliferation of smaller trades aimed at satisfying demand for hats, stockings, caps, and the like, also prospered during the seventeenth century.[9] These novel industries attest to extensive consumer interest in the new ephemeral goods, albeit at a rudimentary level compared to later developments. Nevertheless, fashion evidently motivated popular consumption.

'Fashion', stated Louis XIV, 'is the mirror of history.'[10] It reflects political, social, and economic changes, rather than mere whimsy. At the time of the Restoration in England, fashions were no longer the sole preserve of the nobility and aspiring gentry, but neither were they the inspiration of local majesty. The styles of dress set at court reflected a French pre-eminence throughout Europe. French vogues followed in the train of Charles II as he ascended the English throne, marking the start of a Gallic domination of fashion in Britain that was to last for more than a century. The sumptuous attire of the Sun King was replicated in the English court. A winter wardrobe of heavy velvets, brocades, and the finest embroidered wools did duty when the summer clothing of taffetas, satins, silks, and lace was not in use.[11] All the most gorgeous aspects of French attire were adopted by the court of Charles II during the initial period of reaction to Puritan asceticism. For both the lady and the gentleman, fashionable dress was the glittering reflection of the status of those whose hands wielded authority. 'A nobleman was expected to dress in luxurious garments ... so that both his rivals and his subordinates would interpret correctly these symbols of his wealth and power.'[12] Centuries of sumptuary laws were rooted in that philosophy of dress, linking rank

---

[9] Joan Thirsk, *Economic Policy and Projects* (1979), provides an excellent account of the numerous projects begun during the sixteenth and seventeenth centuries, including stocking-knitting and hat-making, all directed to the requirements of the English market and designed as substitutes for articles previously imported. See also, Thirsk, 'Fantastical folly' and David Corner, 'The London Hatting Trade, 1660–1800', *Textile History* (1991), special issue of papers from the Pasold Conference on the Social and Economic History of Dress, London, 1985.
[10] Quoted in Michael and Ariane Batterberry, *Mirror Mirror* (1977), 145.
[11] Batterberry and Batterberry, 145.
[12] Rosalind K. Marshall, *The Days of Duchesse Anne* (1973), 83.

with 'seemly splendour' and social standing with the leisure to trifle with elegancies of dress.

French fashions evolved in the late seventeenth century to assume a less rigid and more flowing style. This garb first found favour among the younger set at Versailles and then among the European nobility.[13] The construction of the new-style gowns as simple—a great part of its novelty. The configuration of the gown depended almost entirely on the characteristics of the fabric employed in its creation. The preference for lighter, more fluid garments, and the pre-eminence of French modes ensured the sale of French silks. Moreover, fashionability was determined not only by the wearing of a silk gown, but also by sporting the newest pattern. A richly flowered silk was at its zenith for fewer and fewer months as the call for novelty became more pressing among the noble consumers, in an effort to distinguish the *haute monde* from their thrusting imitators.

The cost of a new gown lay almost entirely in the cost of the cloth; it was the fabric used that placed one in the forefront of fashion, or among the *passé*. Thus, when novelty was demanded, it was demanded first from the silk designers. Evolutions in patterns and styles of clothing had long occurred, but slowly, sometimes haphazardly, depending on the movements of merchants or diplomats across Europe for the transmission of the new vogue.[14] By the end of the seventeenth century there was a visible change in the pace of the fashion cycle, as silk patterns were developed and fashions unfolded at greater speed. A contemporary writer explained how, for 'many years past the manufacturers of silks have puzzled both their own and tortured the pattern-drawers brains to contrive new fashions'.[15] The word '*nouveau*' became the most desirable appellation to be attached to a length of dress silk, assuring the buyer the cachet of fashionability when the garment was worn in public. So intense was the passion of those following the Parisian modes, that in 1678 the Paris fashion journal *Mercure Galant* promised its readers a special supplement describing and depicting the new spring designs for silk.[16] This marked a new stage in the formalized dissemination of fashions.

---

[13] François Boucher, *20,000 Years of Fashion* (1967), 261.

[14] This phenomenon is clearly described in Thirsk, 'Fantastical Folly', 54–5.

[15] G. Smith, *Laboratory, or School of Arts* (1756), quoted in Peter Thorton, *Baroque and Rococo Silks* (1965), 19.

[16] Thornton, 20–1. Discussion of the later influence of English women's magazines on the dissemination of fashion can be found in Chapter 5 below.

The more rapid alterations in fabrics influenced not only France but all those trades which were linked to the production of associated goods. Joan Thirsk writes that the 'vagaries of fashion disturbed the ultra-sensitive moralists of the day; but they also disturbed the politicians, for every change of style in dress had economic consequences, and these could be damaging to native industries'.[17] The rage for French silks, for vivid, stylized, floral prints in softly flowing materials, burned in the hearts of English gentry too, and even in the lower echelons. One English commentator wrote scathingly of the mania, noting that 'the laudable English fashions of former times began to alter in favour of France. The women's hats were turned into hoods whereby every maidservant in England became a standing revenue to the French king of the half of her wages.'[18] The insatiable demand for Gallic modes stimulated not only the production of elaborately flowered silks for court dress, but all the cheaper imitative fabrics. Those who saw the latest fashions on the streets and in the corridors determined to copy as much of the observed style as they were able. Such fashions transplanted to England moved slowly down the social scale, and in their diffusion were ultimately adapted to the needs of the wearer.

Clothing styles conspired to aid the democratization of fashions, through a suggestive display of underclothes, offset by the new, informal, softly flowing gown known by a variety of names, such as mantua, nightgown, *battante*, *négligé*, *robe de chambre*, or *sach*. Such gowns were worn more and more on occasions where formal dress was not required, as well as by that greater number for whom formal dress was never necessary.[19] The chemise now slid half off the shoulder, no longer concealed by the gown, while the petticoats were openly displayed with the overskirt looped back. The mantua, loosely cut, was pleated at the shoulders and tied at the waist.

This style of dress echoed in the realm of fashion the profound economic changes that were already moving English society away from its traditional past and towards a new order. Up until this period, although the clothing of the élite had been elaborate, there had been essentially only two or three sorts of apparel adapted for all requirements, most of the components of which placed it above what could be aspired to by those of lower social and economic status. As

---

[17] Thirsk, 'Fantastical Folly', 51.
[18] Quoted in P. J. Thomas, *Mercantilism and East India Trade* (1926), 25.
[19] Boucher, 261.

articles of apparel were simplified, emulation of the dress became more feasible. Stockings were a case in point. Once knitted hose became commonplace more varieties and permutations on the basic shape were produced and could be readily copied by the baser disciples of fashion. Peter Stubbs considered that 'no people in the world are so curious in new fangles as they of England be'.[20] Whether or not that was the case, men and women looked to those above them for an interpretation of style which they themselves could then follow, hoping to present in their dress the appearance of a higher status. So popular were the clocked, coloured, embroidered, and patterned stockings with the lower ranks during the mid-sixteenth century that informers were placed at the gates of London to watch for those wearing hose incompatible with their station.[21] But though these 'sadde and discrete personages' examined passers-by diligently for the offending 'great and monstrous hosen, silk, velvet or weapons restreyned and prohibited', legislators were as unsuccessful in limiting the spread of fashionable clothing through society in London as they were elsewhere in England. All legal prohibitions were removed from the statute books in 1604. However, this did not reflect official sanction of the diffusion of fashions into the lower ranks.[22] Latimer had voiced the common complaint prior to the repeal, bemoaning the fact that, although 'There be lawes made and certain statutes, how every one in his estate shall be apparelled ... God knoweth the statutes are not put in execution'.[23] With no legal prohibition of any sort and more easily replicable styles, England abounded with men and women mimicking the newest modes as best they could.

Conservatives decried the proliferation of informal patterns of dress. The Duchesse D'Orleans considered them distasteful, writing in 1695: 'I don't know why people have so many different styles of dress; I wear Court dress (*le grand habit*) and a riding habit; no other; I have never worn a *robe de chambre* nor a mantua; and have only one *robe de nuit* for getting up in the morning and going to bed at night'.[24] But for

[20] P. Stubbs, *Anatomie of Abuses* (1595), 10, quoted in Thirsk, 'Fantastical Folly', 51.
[21] Ibid.
[22] In fact no agreement could be reached with regard to the creation of new sumptuary legislation. A full examination of this question is presented in N. B. Harte, 'State Control of Dress and Social Change in Pre-Industrial England' in D. C. Coleman and A. H. John (eds.), *Trade, Government and Economy in Pre-Industrial England* (1976).
[23] Wilfrid Hooper, 'Tudor Sumptuary Laws', *English Historical Review*, 30 (1915), 436–8, 443, 447–8.
[24] Duchesse D'Orleans, *Correspondence* (1695), quoted in Norah Waugh, *The Cut of Women's Clothes, 1600–1930* (1968), 112.

the growing middle section of society, particularly those who lived in or visited the city, the absence of heavy velvets, fur trim, and lavishly jewelled gowns in the fashionable dress of the day meant that they could and did attempt to approximate the current modes. Mrs Aphra Behn wrote condescendingly of 'Your frugal huswifery Miss in the Pitt at a Play, in a long scarf and Nightgown'.[25] But opposition by the more orthodox factions could not halt the amendments to fashion, nor could they restrict their usage to the hereditary leaders of society.[26] Once adopted, the mantua remained one of several standard dresses, worn in all ranks.

The gentlemen were not loath to follow the lead of the ladies in sporting less rigid garb, and the shirt assumed an important function as a visible accessory, simultaneously indicating rank and status.[27] One of the scions of the Verney family wrote of this new fashion in 1688, stating: 'I hope you will consider to buy me some good shirts or else some sort of wastecoat, for it is not fashionable for any gentleman to go buttoned up either winter or summer.'[28] The progenitors of this latest style adopted wider sleeves and more opulent cuffs; some discarded the waistcoat altogether and sported their shirts blousing out at the waist, with the cravat carelessly tucked into the buttonhole. One wit described a similar style in 'a fat fellow whom I long remarked, wearing his breast open in the midst of winter . . . A sincere heart had not made half so many conquests as an open waistcoat'.[29]

Once these light, comfortable textiles were worn, the qualities of the materials quickly captured an ever-increasing portion of the market. In the years immediately after the Restoration no domestic manufacturers produced materials that could rival the French imports, much to the distress of the economic nationalists. In 1666, during a period of hostilities with France, Charles II challenged the domination of French styles by championing an eastern form of dress, promoting England's trade with the Orient at the same time as he thumbed his

[25] Quoted in Geoffrey Squire, *Dress, Art and Society, 1560–1970* (1974), 106. 'Nightgown' refers to a type of informal gown.

[26] The duchesse D'Orleans subsequently lost interest in the new sorts of dress, as can be seen from her writings in 1721. 'I only follow the fashions from afar and there are some which I will have nothing to do with, like paniers which I won't wear and *robe battantes* which I detest. I find them indecent and will not allow them in my presence', quoted in Waugh, 115.

[27] C. W. and P. Cunnington, *The History of Underclothes* (1951), 53.

[28] *Verney Memoirs*, quoted in Cunnington and Cunnington, 56.

[29] 'The Levellers, A Dialogue between two young Ladies concerning Matrimony', *Harleian Miscellany*, quoted in Squire, 110.

nose at his illustrious cousin Louis XIV. John Evelyn recorded the symbolic event when:

his Majesties putting himself solemnly into the Eastern fashion of Vest, changing doublet, stiff Collar, bands and Cloake etc: into a comely Vest, after the Persian mode . . . resolving never to alter it, and to leave the French mode, which had hitherto obtain'd to our great expense and reproach: upon which diverse Courtiers and Gent: gave his Ma[jesty] gold, by way of Wager, that he would not persist in his resolution.[30]

Samuel Pepys registered the birth of this politically inspired garb as well, writing that, 'I did see several persons of the House of Lords, and Commons too, great courtiers, who are in it [the Vest] . . . I wish the King may keep it, for it is a very fine and handsome garment'.[31] Pepys later ordered a vest for himself, not wanting to be out of style; while Evelyn had already adopted one, since 'his Majesties had brought the whole Court to it'.[32]

Lord Halifax considered this event not only an advance in fashion, but also politically astute, whereby 'we might look the more like a distinct People, and not be under the Servility of Imitation'.[33] Halifax accounted this incident of some symbolic importance, and he related one of the measures used by the French to end the boycott of French fabrics and fashions. The Duchesse D'Orleans was sent over to England to employ her considerable social cachet to help end the royal boycott, so much to the detriment of France.

It was thought that one of the Instructions that Madame brought along with her, was to laugh us out of these Vests; which she perform'd so effectually, that in a moment, like so many Footmen, who had quitted their Master's Livery, we took it again and return'd to our old Service. So that the very Time of doing this gave a very critical Advantage to France, since it look'd like an Evidence of returning to their Interests, as well as to their Fashions.[34]

The English were made to feel the ire of the French king. Pepys wrote excitedly of the news brought from Paris by Lady Fanshaw, 'the news how the King of France hath, in defiance to the King of England, caused all his footmen to be put into Vests, and that the noblemen of France will do the like'. Such a heavy-handed rebuttal by Louis

---

[30] John Bowle (ed.), *The Diary of John Evelyn* (1985), 216.
[31] Robert Latham and William Matthews (eds.), *The Diary of Samuel Pepys* (1972), vii, 324. [32] *Pepys*, vii, 353; *Evelyn*, 217.
[33] Quoted in John Dennis, *An Essay on Publick Spirit, etc.* (1711), 6.
[34] Ibid. 7.

perhaps signified his determination to defend his nation's hegemony over fashion and its components, as well as forcing compliance from his royal pensioner.[35] Pepys decried the method used to assert French pre-eminence, calling it 'the greatest indignity ever done by one prince to another', adding that this action 'would incite a stone to be revenged'.[36] There is no record of Charles being so incited. Louis could deride his fellow monarch; however, he could not determine the style and clothing that would be worn, no matter how great his temporal power. Furthermore, the nationalist zeal unleashed in this sartorial contest had unforeseen implication for East India traders. Hereafter, the products of India would be associated for some with the efforts of the English monarch to establish independent standards of dress, ultimately to the commercial benefit of the merchants in oriental textiles. The court's public endorsement of these products brought them to the attention of the fashionable throughout the nation.

THE CALICO CRAZE

In company with several other European monarchs Elizabeth I had granted a trading monopoly to a company of merchants, eager to hazard the dangers of oceans, pirates, and competing traders, for the chance of riches. At its founding in 1600 the East India Company received 'large Privileges for the Improvement of Navigation, the Glory of the Kingdom, and the Increase of Trade'.[37] High hopes for the successful establishment of a powerful English presence in the spice trade were dashed within a few decades, when the English were driven from the Indonesian Archipelago by the Dutch, settling instead in several coastal ports on the Indian subcontinent. But during the seventeenth century a humiliating defeat was turned to serendipitous advantage. India was a cornucopia of textiles, in colours, patterns, and qualities unrivalled in Europe. Here were unanticipated opportunities for profitable trade. Samples of Indian textiles arrived in London early in the century. Small quantities were dispatched initially for their curiosity value, but were soon sold for a handsome profit. The

---

[35] Charles II's willingness to accept a pension from Louis XIV is well known. At least £300,000 was given Charles in return for favourable political decisions. Demonstrations such as the one in 1666 might well have raised his price, though it would not have endeared Charles to his autocratic cousin. [36] *Pepys*, vii, 379.

[37] Sir Dudley Diggs, *Of the East-India Company, its first Erection and Progress* ... (1614), in John Smith (ed.), *Chronicum rusticum-commerciale* (1747, repr. 1968), i, 137.

Company quickly took advantage of the fancy for oriental materials of unique design; they could not afford to ignore the huge potential market inherent in fashionable London society.[38] Presently, East India Company officials were exhorting their representatives in India to ship out specific cottons that would appeal to English tastes.

The Quilts of chints being novelties produced from £5,5s,0d to £6 the pair, a further supply therefore desired, and both as regards those and the Chintz, more should be made on the white grounds, and the branches and flowers to be in collors, and not to be (as these last sent) all in general of deep red ground and sadder collors.[39]

The East India Company launched an aggressive and determined promotion of Indian cottons. In part this required that the goods made in India conform to the fashions in Europe, while still retaining that aura of oriental exoticism. A fine line separated the novel and attractive from goods too excessively unusual. In 1645, directions arriving at the factory in Madras recommended that regional fabric motifs be abandoned and those of the Gujarat area be used instead, because of the latter's popularity with buyers in England.[40] British society was pleased to discern an elegance and attractive novelty in the Indian fabrics. Where king and court adopted oriental textiles for their occasional wear, the lower social ranks were only too ready to follow.

Samuel Pepys was typical of the ambitious, middling sort of man, eager to be seen as *à la mode* both by his superiors and inferiors, determined that even visiting family members dress in a manner that gave him credit.[41] The cost of the East Indian calicoes, chintzes, and muslins allowed even the less affluent to own vivid, floral patterned, checked, or plaid clothing or soft furnishings. Moreover, many of the fabrics could be substituted for costly French silks. In 1663 Pepys, his friend Creed, and Pepy's wife Elizabeth went to Cornhill, one of the most fashionable shopping districts of London, where, 'after many

---

[38] Thomas, 25–6; K. N. Chaudhuri, *The Trading World of Asia and the English East India Company* (1978); the latter provides the most detailed account of the process of trade expansion in Asia by the English East India Company.

[39] Letter Book, 1641, quoted in G. P. Baker, *Calico Printing and Painting in the East Indies in the XVIIth and XVIIIth Centuries* (1921), 30.

[40] John Irwin, 'Indian Textiles in the Seventeenth Century' in John Irwin and P. R. Schwartz (eds.), *Studies in Indo-European Textiles* (1966), 30.

[41] On 18 February, 1666, Pepys notes in his diary a conversation with his wife wherein he discussed his sister Pauline's impending visit and his concern for her appearance.

tryalls bought by wife a Chinke [chintz]; that is, a paynted Indian Callico for to line her new Study, which is very pretty'. The Indian cloth must have been a success with Elizabeth, for two months after the initial purchase we find Creed sending Pepys a gift, 'a very noble parti-coloured Indian gowne for my wife'.[42] Pepys congratulated himself on his growing prosperity and the clothing which he could now afford to buy for himself and his wife. Such attire was immensely important to him during the course of his career, a conceit which he readily admitted in the pages of his diary. By 1663 he had already advanced sufficiently to be able to dress himself and his wife in a suitable manner, consistent with his advance in position.

I hope I shall not now need to lay out more money a great while, I having laid out in clothes for myself and wife . . . these two months . . . above 110 *l*. But I hope I shall with more comfort labour to get more, and with better successe then when, for want of clothes, I was forced to sneak like a beggar.[43]

Pepys knew, long before Mandeville wrote the words, that:

People, where they are not known, are generally honour'd according to their Clothes and other Accoutrements . . . It is this which encourages every Body . . . to wear Clothes above his Rank, especially in large and populous Cities, where obscure Men may hourly meet with fifty Strangers . . . and consequently have the Pleasure of being esteem'd . . . not as what they are, but what they appear to be.[44]

By 1680 many fabrics of European origin were being supplanted by the new Indian textiles. Furthermore, the uses to which Indian materials were put altered as wider uses were found for them and outlets sought in all social ranks. Members of the middling and lower orders first began to buy the vibrant cottons as a cheap facsimile of the brocades and flowered silks favoured by the aristocracy. The East India Company observed the initial reluctance of the wealthier classes to use their imports for clothing, and set out to overcome their resistance. Sir Josiah Child was the author of the policy and he, along with the other directors of the Company, used their connections with friends and acquaintances to champion Indian textiles. 'Favours and services were extended to prominent and useful people; samples of Indian goods were distributed where they would attract attention and

---

[42] *Pepys*, iv, 299, 391.
[43] *Pepys*, iv, 358.
[44] Bernard Mandeville, *The Fable of the Bees: Or Private Vices, Publick Benefits* (1714, repr. 1924), 127–8.

create a fashionable demand.'[45] Indian manufactures had to be perceived as a desirable commodity by the gentry and middling ranks, to extend sales and confer a cachet to these goods. Thus, Charles II was given 'voluntary contributions' between 1660 and 1683 of £324,150 by the Company and, in return, was pleased to be seen in an oriental-style waistcoat, confirming the desirability of Indian fabrics to all aspirants of fashion.[46]

In 1664 over a quarter of a million pieces of calico were imported into England, equalling 73 per cent of the Company's trade.[47] About one-third of this total was bought in England, by those anxious to be 'esteem'd by a vast Majority'. Men and women wanted to appear in comfortable circumstances, respectable, prosperous, and in fashion, and none more so than Samuel Pepys. When he paid John Hayls to paint his portrait Pepys arranged to be wearing an Indian silk morning gown, or banyan, a piece of apparel that was *de rigueur* among the London gentry and nobility, but which he was obliged to rent in order that the completed portrait would commemorate him in the style of the *haute monde*.[48] Whether or not Pepys could afford to own such an item of clothing, he intended that his portrait would show a fashionable attire modelled on that of his superiors.

The second prong of the campaign devised by Child was directed at serving the home market with the quality and types of merchandise most in demand. Excellence in manufacture and originality of design concerned the East India Company above all else. Only such a combination could ensure an extension of sales among the more discriminating buyers in the higher ranks, and their patronage in turn determined successful acceptance by the imitative middling ranks. The instructions set out in a 1683 letter from London to the Surat factory illustrate the attention paid to marketing details.

You did exceedingly well in the observing our direction about Chints . . . Now we hope you have more patterns . . . you cannot imagine what a vast number of them would sell here to content, they being the ware of Gentlewomen in Holland, but of the meaner sort here, for which reason a great part of them must be painted upon better cloth than formerly they used to do. We give you

---

[45] Audrey W. Douglas, 'Cotton Textiles in England: The East India Company's Attempt to Exploit Developments in Fashion, 1660–1721', *Journal of British Studies*, 8: 2 (1969), 29.   [46] Douglas, 29.   [47] Chaudhuri, 282.
[48] *Pepys*, vii, 602. Pepys remarked in his diary that Sir Philip Howard, younger brother of the earl of Carlisle, was wearing a 'night-gown and Turban' when Pepys called upon him: the former being the loose robe of silk worn over shirt, breeches, and hose on informal occasions, the sort of garment Pepys coveted. Ibid. 378.

this early hint that you may make great provision of them beforehand. 200,000 of all sorts in a year will not be too much for this market, if our direction be punctually observed in the providing of them.[49]

Within four years the directors of the Company were pleased to write that Indian calicoes had become 'the Ware of Ladyes of the greatest quality, which they wear on the outside of Gowns and Mantuoes and which they line with velvet and Cloth of Gold'.[50] Others less sanguine when recounting this success, concurred with Defoe when he described the penetration of Indian commodities in these terms: 'it crept into our houses, our closets and bedchambers; curtains, cushion, chairs, and at last beds themselves were nothing but Callicoes or Indian stuffs'. Defoe concluded in disgust that, 'everything that used to be made of wool or silk, relating either to the dress of the women or the furniture of our houses, was supplied by the Indian trade'.[51] The ubiquitous Indian textiles blurred the divisions in the social hierarchy within and without the household, until in the end, as the husbands' common complaint went, it became difficult to know their wives from their chambermaids.

The introduction of Indian cottons into the English market was tentative at first; painted cotton robes were initially worn only before retiring or upon rising. However, the movement of fashion away from stiff, restrictive garb assisted the East India Company's drive into the market. The transformation in the structure and composition of women's dress is described best by Norah Waugh:

The changing style in women's dress that took place towards the end of the seventeenth century was similar and just as revolutionary as that taking place in men's clothing. As the stiff doublet was being replaced by the coat so the heavily-boned bodice was making way for the looser gown. The distinctive style of dress design which evolved from the loose gown was achieved not by cutting, but by draping the material into pleats which molded the figure and then fell into graceful folds into the skirt.[52]

The alterations in fashions in the seventeenth century worked to the advantage of the East India Company, ensuring a strong demand for the brilliantly painted and printed clothing fabrics. Moreover, accessories could also be made of the calico or muslin currently in vogue. Contemporary writers listed the extent to which the Indian materials could be adapted to styles of the day. Decorative aprons joined the

[49] Letter Book, quoted in Irwin, 17.
[50] Despatch Book, 23 Mar. 1687, vol. 91, p. 275, quoted in Chaudhuri, 281.
[51] Daniel Defoe, *Weekly Review*, 31 Jan. 1708.   [52] Waugh, 65.

highly visible petticoats as articles of display, both of which could be made of calico or chintz; head-dresses, hoods, sleeves, nightrails, and pockets were also to be found in Indian cottons. Gentlemen had muslin made into cravats, and they too wore shirts, cuffs, gowns, and carried handkerchiefs of Indian manufacture. Both men and women wore Indian-made stockings.[53] The Company also continued to encourage the production of household goods that might appeal to the middling and lower sections of the market, aiming to stimulate new markets among as wide a cross-section of the population as possible. A 1682 letter directs the factor to have bed-covers, curtains, and bed-hangings made, stipulating that these items be,

ready made up of Several Sorts and Prices, strong, but none too dear, nor any overmean in regard you know our Poorest people in England lye without any Curtains or Vallances and our richest in Damask ... Possibly some of these things may gain that repute here as may give us cause of greater enlargement in them hereafter ...[54]

The assortment of fabrics shipped from India was so extensive that in 1696 an anonymous merchant published a catalogue and description of these and all the other fabrics then available. His purpose was explained as,

Shewing How to Buy all sorts of Linnen and Indian Goods: Wherein is perfect and plain Instructions, for all sorts of Persons, that they may not be deceived in any sort of Linnen they want. Useful for Linnen Drapers, and their Country Chapmen, for Seamstresses, and in general for all persons whatsoever.

The author aimed to enable educated consumers to buy textiles of the quality and characteristics required and so avoid being duped into purchasing damaged cloth or those sorts unsuited to the buyer's purpose. Each variety of fabric was listed under alphabetical headings. The cottons known as Bettilies, for example, were described as some of the best sorts of muslins, ideal for 'Cravats, or Heads for Women, or use for several things, it being a more agreeable wear than any other slight Cloths, and is strongest of all Muslings that comes into England'. The 'Original Bettilies' was thought by the author to be especially suited 'for Cravats for ordinary Tradesmen ... by reason they are not only strong but thick ... usually the cheapest of all those Bettilies I have named'.[55]

[53] Cary, 4–5. [54] Manuscript Letter Book, 1682, quoted in Irwin, 36–7.
[55] J. F., *The Merchant's Warehouse laid open: Or, The Plain Dealing Linen-Draper* (1696), pp. i, 2, 3.

This small volume breaks down the featureless trade statistics of imported East India textiles and reveals exactly how important they were throughout the national market. (See Plate 1, which lists goods sold in one auction.) The author moves through the myriad sorts of Indian fabrics then in use: the range of materials was striking. Barras, for example, was described as a 'very useful Cloth for packing of Goods for the Country and when well whited is good for ordinary Sheets for Poor People and Servants'.[56] Some facts seemed self-evident to the author. Calicoes, for instance, were introduced simply as 'of general use with us', being fabrics that were 'much worn in Shirts and Shifts'.[57] The ubiquitous chintzes were thought to need correspondingly little introduction, 'being of so general use in this City'. The author went on to relate the names of various sorts of chintzes and the probable consumers for each sort. 'Chercanneys' received several lines outlining its features: 'cheickered [sic] with variety of colours, as Red, Yellow, Blew and Green . . . wears very well in anything you shall think fit to use it for, as Lining of Beds, Window Curtains or Morning Gowns, or Under Petty Coats'. Derriband, in contrast was thought the 'worst of any Callico that at present is in use', while dungarees, another sort of calico, was stronger and commonly made into shifts by 'ordinary People'.[58]

Great assortments of fabrics varying in price, colour, pattern, texture and price were readily available. Such a diversity is commonly associated with later industrial production. But in the huge assortment of East India cottons there is evidence of the great choice available from pedlars and shopkeepers long before the industrial age. Joan Thirsk identified a tremendous collection of inexpensive products manufactured outside guild control in late sixteenth- and seventeenth-century England, writing that: 'When we survey the magnificent range of choice available to the customer in seventeenth-century England, we are compelled to think deeply about the economic significance of quality and variety in consumer goods, and the influence which different classes of customers exerted upon producers.'[59] The augmentation of consumer goods not only provided work, but also gave to the labourers in the domestic system, and others associated with the trades, extra income with which they could purchase stockings, ribbons, buckles, and buttons to improve their own appearance.[60]

[56] J. F., *The Merchant's Warehouse laid open: Or, The Plain Dealing Linen-Draper* (1696), pp. 3, 4.   [57] Ibid. 4.   [58] Ibid. 7, 14.
[59] Thirsk, *Policies and Projects*, 107.   [60] Ibid. 107–32.

Similarly, the East India Company directed its products at the huge markets presented by the middling and lower ranks, as well as cultivating the higher-quality products for the élites. This policy was the essence of their success in the English domestic market; it was the policy built on the new consumerism extant in England.

Sir Josiah Child set out specifically to develop a market for cottons where none had previously existed. The extent to which calicoes pervaded the domestic market is evident from a 1688 inventory of a Preston draper. Recorded were 'white buckram calico at under a shilling a yard, white calico, printed and glazed calico at 1s. 1d., brown calico at 10d., black, blue and "coloured" calico at 11d., broad glazed calico at 1s., stained calico at 1s. 2d., and 1s., narrow flowered calico at 9d., and, finally, coloured calico at 1s. 7½d'.[61] Chapmen, pedlars and hawkers carried a range of oriental textiles for their customers in addition to their other merchandise. John Smyth of Randwick, Gloucester, had among his stock on his death in 1691 '10 yards of Callicow at 6d. yard', plus '31 yards of Callicow in Remnants' valued at 15s. 6d. Thomas Simpson of Caythorpe in Lincolnshire included '2 yard quarter blew Caleco', as well as four varieties of muslin and three of calico in his 1714 inventory. A Kent-based petty chapman named John Cunningham, who died in 1690, had two pieces of calico listed among his effects, plus three types of muslin; while Walter Martin of York carried three sorts of muslin and one of calico in 1712. The latter conducted his trade on foot out from York; Cunningham and Simpson were mounted chapman and stall-keeper, respectively; while Smyth was a shopkeeping chapman of some means.[62] All of these men were small and middling tradesmen, part of an extensive chain of retail distribution that carried various sorts of Indian cottons throughout Britain, both as yard-goods and as ready-made accessories, such as the muslin neckcloths sold by John Cunningham for 9d. each.[63] The publication of *The Merchant's Warehouse laid open* in 1696, and the surviving inventories confirm the substantial demand for these goods beyond the boundaries of London, penetrating rural England, lowland Scotland, and Wales, along the pedlars' trails and post-roads.

[61] Wadsworth and Mann, 124.
[62] Margaret Spufford, *The Great Reclothing of Rural England* (1983), 156–7, 163, 169–70, 190–3.
[63] Ibid. 103. The *Journal of the House of Commons* listed an intriguing array of ready-made items that were imported into London during 1698. Among the articles listed were calico nightgowns—6, petticoats—11, shirts—4,112, waistcoats—39, and neckcloths—49,374. *Journal of the House of Commons*, xiii, 176.

The establishment of these new decencies among the middling orders and the fostering of new wants among the labouring people threatened accepted views and precipitated a crisis. Some opponents directed their attacks at the East India Company and the new markets they had acquired; equally vocal opponents decried the greater commercial activity and excessive consumerism exhibited by segments of the populace. The critics were answered. Sir Dudley North, for example, made plain his support for this more vigorous domestic trade, noting that the, 'main spur to Trade, or rather to Industry and Ingenuity, is the exorbitant Appetites of Men, which they will take pains to gratifie, and so be disposed to work, when nothing else will incline them to it; for did Men content themselves with bare Necessaries, we should have a poor World'.[64] As the tempo of the debate quickened the battle-lines became more sharply defined. The unprecedented rate of sale of exotic foreign textiles enraged wool manufacturers, landowners, and weavers; in their eyes all East India textiles intruded into their home market, in defiance of settled institutions. To others, the popularity of calicoes among almost all ranks of the people blurred the boundaries in the social hierarchy to an intolerable degree, to the detriment of the health, order, and prosperity of the kingdom. Opponents of the trade united in defence of traditional industries, established interests, and existing hierarchies.

Confirmation of incipient social disorder came from many quarters. A common sight to outrage the opponent of the calico craze was described by Bernard Mandeville. He advised anyone who 'takes delight in viewing the various scenes of low Life, may on Easter, Whitsun, and other great Holidays, meet with scores of People, especially women, of almost the lowest rank, that wear good and fashionable Clothes'.[65] To many this was an indictment of a society grown lax, of order distorted through the introduction of frills and extravagances into 'almost the lowest rank' of society. The demands of the middling and even some of the labouring classes for new goods continued unabated, bedevilling many, where even:

> The Cinder Wench, and Oyster Drab,
> With Nell the Cook and hawking Bab,
> Must have their Pinners brought from France,
> Appear most gay and learn to Dance.[66]

[64] Sir Dudley North, *Discourses upon Trade* (1691), quoted in Appleby, 505.
[65] Mandeville, 128.
[66] Daniel Defoe, *The London Ladies Dressing-Room* (1725), 7.

The vocal insistence of the wool and silk industries that they be protected continued with the growing chorus of critics demanding voluntary or enforced temperance of English men and women. The calico campaign attempted to shore-up traditional industries at the same time as it sought to reimpose sumptuary restraints on the populace and entrench the aristocratic preserve of fashion.

THE CALICO CAMPAIGN: THE GATHERING FORCES

To the determined adherents of mercantile theory the commercial transactions with the Orient had nothing to recommend them. It was a trade that, in Cary's view,

> exports little or none of our products or Manufacturers, nor supplies us with things necessary to promote Manufactures at home, or carry on trade abroad, . . . its Imports hinder the consumption of our own Manufactures, and more especially when those Imports are chiefly purchased of our *Bullion* or *Treasure*.[67]

In addition, the wool trade suffered intolerably. Britons had always worn woollens; wool fabric of one kind or another was an accepted garb which sustained craftsmen and landowners alike. In the opinion of many, two-thirds of society ought to wear only the 'Serges, Tammies, and Norwich Stuffs' of domestic making, to ensure the prosperity of the kingdom.[68] John Cary tried to explain 'why the People of England are so much against their native Manufactures, as to be more in love with *Calicoes* and *Indian Silks*'. He found no acceptable explanation for this preference, concluding that the 'chief Reason' being 'Fashion and Imitation of one another, though many others are alleged, as the Ruffness and ill Colour of Woollens, which keeps it from answering the ends of *Calicoes*'. Cary firmly asserted that these claims 'are not substantial but pretended Reasons'.[69]

The onslaught against the East India Company and their chief import continued with unabated vituperativeness throughout the last decade of the seventeenth century. The barrage of abuse grew in proportion to the popularity of the Indian fabrics in the home market. England's involvement in the trade with Asia sparked some of the most contentious debates of the time, resulting in a legacy of literature on

[67] Cary, 2, 7. [68] Houghton, 2.
[69] Cary, 6–7.

economic theory and social practice.[70] There was no precedent for the East India trade. The centuries of trade between England and Europe advanced no paradigm against which the situation could be judged; contemporary commercial enterprises in the Americas and Africa did not replicate the trading pattern with the Far East; a new dilemma was introduced. Trade with the Orient brought enormous wealth to members of the East India Company. But was this wealth gained at the expense of the other members of society, as some critics suggested? The historic economic order dictated protection for the wool industry and, by implication, for the pastoralist landowners, while the imperatives of the new-born commercial enterprise appeared to require untrammelled access to the home market and freedom from government restraints.

England was in a transitional period, in which a volatile commercial expansion in trade had been stimulated by dramatic domestic consumption. This process had been sanctioned with regard to sugar and tobacco, as no vested interests were threatened, and all the refining processes could take place in Britain. Such was not the case with the East India trade. Indian cottons became the focal point of two opposing forces; government had to attempt to restore an equitable balance, weighing the relative merits of each case.

Sir Josiah Child cultivated close, cordial ties with the court for strategic reasons. He made certain that those in power and those with access to the ears of the powerful should remain sympathetic. In 1680, during one of the early parliamentary debates on the trade, William Love railed against Child's practices and Company activities. In addition to the Company's distribution of lavish 'New Year's gifts', Love recounted how Child enlisted influential men 'into the company . . . chosing them of the Committees, though they understood no more of trade than I of physic'.[71] As well, Company servants swore an oath of secrecy, successfully hiding the more intimate details of business behind this screen of silence, so, as Love remarked, 'there is no Way left to reach them'.[72] A more literary critic portrayed other instances of the Company's efforts to curry favour among those

---

[70] William J. Barber summarizes some of this material in *British Economic Thought and India* (1975). A contemporary anthology of tracts, pamphlets, and letters on the East India trade, with extensive editorial comments in the footnotes, can be found in John Smith (ed.), *Chronicum rusticum commerciale*, vols. i and ii (1747).

[71] William Love, *Debates in Parliament: 1689, the 9th of November*, in *Chronicum*, i, 354.   [72] Ibid.

habitués of the court with official and unofficial connections to the powerful, through the offering of

> ... great gifts of finest touches
> to lordes and ladyes, Dukes and Duchess
>
> . . . .
>
> By pouring gold in plenteous showers,
> In ladyes laps that bore great powers,
> Which strangely altered all their measures
> Such charms there are in hidden treasures.[73]

For a time such alliances checked the efforts of opponents to disrupt the trade. But as the health of the wool industry deteriorated the East India Company faced stiffer attacks.

The English wool trade was in a generally unsettled state. The switch in popular tastes in favour of the New Draperies has been well documented. During the seventeenth century the entire system of production had been shaken and reformed under the pressure to produce lighter materials, while the traditional woollen manufacturing districts suffered a series of shocks, from the Cockayne débâcle to the continuing diminution of their European markets. The European cloth market had entered a period of dislocation and innovation, from which it would emerge profoundly altered. Clothiers from England and the Netherlands competed for markets for their worsted, camlets, and says in the Mediterranean regions, while the appearance of rival wool manufacturers in Spain further tested England's position in the international cloth trade.[74] These alterations in the structure and requirements of the European markets were further exacerbated by the intermittent wars, but no one associated with the production and sale of wool textiles was prepared to accept any other rationale for the falling off of trade than the popularity of Indian textiles. Wool merchants, clothiers, weavers, and others connected to England's staple were in agreement on that point.

By the best computation that can be made, we now spend in this kingdom per annum to the value of two or three hundred thousand pounds worth of goods manufactured in East India: what part there of are spent instead of our stuff, serges, cheneys and other goods, I have leave to every man's judgement that hath observed how their Persian silks, Bengals, printed and painted calicoes, and other sorts are used for beds, hangings of rooms, and vestments of all

---

[73] *Prince Butler's Case* (1696), quoted in Thomas, 64–5.
[74] Charles Wilson, 'Cloth Production and International Competition in the Seventeenth Century', *Economic History Review*, 2nd series, 13: 2 (1960), 213–16.

sorts. And those goods from India do not only hinder the expense of our woollen goods by serving instead of them here, but also by hindering the consumption of them in other parts too, to which we export them.[75]

Questions raised in the House of Commons about the spread of East India wares were followed in the 1680s by stronger assertions as to the rights of the wool industry. There was general concern in England about the measures taken by various European governments to promote locally produced woollens, and Parliament sought to counter foreign mercantile restrictions and slipping home sales. In a statement with many modern echoes, John Basset asserted that the East India trade had to be regulated and restrained as a prerequisite to any firm solution of the problems of the wool industry. English wool fabrics could not be made as cheaply as the Indian materials, and therefore were at a disadvantage when both products were competing for the same market; no difference in the quality or the function of these disparate textiles was recognized by Basset in his argument. He went on to stigmatize the whole eastern trade as one that did

us much Prejudice, as must, in the End, be the Destruction of our Manufacturing Trade ... if not looked after; and the more likely, because People in India are such slaves, as to work for less than a Penny a Day; whereas ours here will not work under a Shilling; and they have all materials also very reasonable, and are thereby enabled to make their Goods so Cheap, as it will be impossible for our People here to contend with them.[76]

As the sale of English wool cloth fell off in foreign markets, a discernible slump was also experienced at home. Parliament felt compelled to defend this staple industry. Earlier steps had already been taken to assure a continued market for wool fabrics with the passage of an Act that required the dead to be buried in English wool. Parish officials had to assure compliance with the terms of this Act; thus from 1678 onwards parish records carry the notation 'buried in wool' in accordance with the new regulations.[77] Initial success with a measure such as this inspired Parliament to greater heights of legislative inventiveness, attempting to reintroduce elements of sumptuary legislation for the living, as they had been so successful with

---

[75] Sir John Pankhurst, *Debates in Parliament: 1680, the 9th of November*, reprinted in *Chronicum*, i, 351.   [76] Basset, *Debates in Chronicum*, i, 351.
[77] The Act stipulated that: 'No corpse of any person ... shall be buried in any shirt, shift, sheet or shroud ... made or mingled with flax, hemp, silk, gold, or silver, or in any other than what is made of sheep's wool only.'

the dead. In 1689 the Commons attempted, to no avail, to pass an Act that would enforce the wearing of wool clothing six months of the year. Even the vagaries of the English climate did not make this proposal palatable. Ten years later, when years of European wars had further disrupted the principal markets for English wool textiles, the Commons focused on 'all magistrates, judges, students of the Universities, and all professors of the common and civil law' in the hopes that they could be required to 'wear gowns made of the woollen manufacture' all year round.[78] When neither of these attempts succeeded, the lower House turned to female servants as consumers who might be coerced into wearing wool, although, given the fondness for fashionable clothing among female servants, this seems a rather vain attempt at sartorial compulsion. A bill was narrowly defeated that would have compelled female servants in Britain and the colonies, paid £5 a year or less, to wear felt hats of English manufacture.[79] The aim of these frustrated sumptuary initiatives, it was claimed, was to ensure that 'both the living and Dead... be wrapt in woollen; indeed no other law is wanted to complete the business but only one, that our perukes should be made of wool'.[80] Under the spur of the wool industry Parliament created a limited market for woollens among the dead, but they had not yet found a way of curbing the new consuming urges among Britons not yet in that powerless position. Restrictions in dress could be enforced only with great difficulty; legislative restraints had been lifted nearly a century earlier, leaving choice and cost as the great restraining factors in dress.

Following the Glorious Revolution, the old East India Company spent time, energy, and money to exclude encroaching merchants and protect its monopoly privileges. Some of the influence of the East India Company was eroded with the coming of William and Mary to the throne. Envious merchants with Whig connections, excluded from the prosperous oriental trade, determined to end the monopoly privileges of the Company. The challenge from the disgruntled Whig merchants seemed a more substantial menace than did that from the cavilling clique representing the wool interests. Distracted by the chorus of opponents demanding a part of the oriental trade, the Company

---

[78] *House of Lords MSS*, 1689–90; *House of Lords Journals*, xiv, 311, 316; *House of Commons Journals*, xii, 67, quoted in E. Lipson, *The Economic History of England*, 6th edn. (1961), iii, 46.

[79] *House of Commons Journals*, xii, 613, quoted in Lipson, 47.

[80] n.a., *The Advantages of the East India Trade to England* ... (1697–8), repr. in *Chronicum*, i. 420.

expended most of its energies countering charges of bribery and corruption and trying to exclude others from demanding a part of the oriental trade.[81] But by 1698, following a protracted struggle, the old Company was forced to accept the incorporation of new East India Company trading in an area that was once theirs alone.

At the same time as commercial and financial concerns preoccupied Parliament and London merchants, the lesser gentry and small landowners seethed at the imagined neglect they were suffering at the hands of a government which showed neither proper care nor respect for the traditional bulwark of society. In the view of many country Tories the foundations of the kingdom were being overturned; for while parvenu London merchants of doubtful ancestry became richer, the ancient families of England suffered the impoverishment of their estates, honest English artisans starved without work, and ancient manufactures were tossed aside by improvident men and women in pursuit of fashion. The nation had to be reawakened to its true interests. To this end a concerted campaign began, joined by squires and gentry, who equated the prominence of the commercial interest with the decline of their political power. The 'fall of the price of necessities' that put greater purchasing power at the disposal of some, hit hard at those who depended on the sale of agricultural commodities.[82] The sharp and rancorous hostility felt by this country group was directed towards the commercial and financial interests in London who seemed to thrive at the expense of the backbone of England, the country squire.[83] In the 1690s the financial burden of the war effort fell even more heavily on the landowners through the reorganized Land Tax. As their situation became more desperate, their castigation of the London clique blamed for their present state of distress reached greater heights of invective.[84] The eastern trade loomed large as a villain. To the Tory squire, the merchants who traded in Oriental textiles drained the nation of gold, thus making loans hard to come by, while trading in heathen lands and importing 'effeminate luxuries' which undercut wool textiles and impoverished the countryside. The wool trade had campaigned intermittently and unsuccessfully against the importation of Indian textiles for forty years.

---

[81] T. B. Macaulay, *A History of England* (1848, repr. 1960), iii, 502–7; Henry Horwitz, *Parliament, Policy and Politics in the Reign of William III* (1977), 148–52.
[82] Charles Wilson, *England's Apprenticeship* (1965), 186.
[83] J. R. Jones, *Country and Court* (1978), 71.
[84] Ibid. 308–9.

The addition of English landowners to the opponents of the East India trade radically altered the balance of power in this struggle.

Enemies of the East India trade did not accept that the nation would profit through the re-export and resale of Indian goods. A spokesman of this view, John Cary, insisted that, 'there must be a difference made between a Nations growing Rich, and particular Mens doing so ... if we Export the true Riches of the Nation, for that which we consume on our Luxury, tho' private Men may get [rich] by each other, yet the Wealth of the Nation is not any way encreased'.[85] Cary and many others insisted that economic advantage could not result from commercial endeavour unrelated to domestic industry. 'I take the true Profits of this Kingdom to consist in that which is produced from Earth, Sea, and Labour, and such are all our Growth and Manufactures'.[86] The tempo of petitions from wool merchants and others quickened. They insisted that the cause of their decline could be traced directly to the growing trade in Indian textiles, and they demanded that steps be taken to curb the pernicious Indian imports. Even Sir Josiah Child could not deny completely the basis of the complaints by the wool merchants, since the cheaper calicoes had been substituted for light wool mixtures. 'I do admit that the wearing of so many printed calicos had been a prejudice to the complainants and do wish a means might be found to prevent it'.[87] However, at no time would be accept suggestions that the eastern trade should be sacrificed simply to placate the proponents of wool.

Throughout the lengthy debate Child reiterated again and again that the finer muslins and chintzes replaced European imports, and thus England gained by supplying the light textiles demanded by the nation at a lower cost than would be possible with fabrics shipped from the continent.[88] The economic writer Charles Davenant expressed the same sentiments when he wrote that 'a wise State must consider, which Way the Folly of their People can be supplied at the cheapest Rate'.[89] Since the people would always buy novel goods, deemed luxuries by some, Davenant felt that England should encourage imports from the cheapest source, in this case the East Indies. Davenant advised that the government spend its time promoting the

[85] Cary, 2.   [86] Ibid. 3.
[87] Child, quoted by Thomas, 51.
[88] Sir Josiah Child, *The East-India Trade the Most National* ... (1681), repr. in *Chronicum*, i, 358.
[89] Charles Davenant, *An Essay on the East-India Trade In a Letter*... (1696–7), repr. in *Chronicum*, i, 411.

sale of wool manufactures abroad, since the comparatively costly English commodity would provide a larger profit when sold overseas than could be had from the cheaper Indian re-exports. Moreover any prohibition, opined Davenant, would do nothing to encourage foreign consumption of English woollens abroad.

For if [the Indian] cottons hinder the Consumption of the Woollen Manufacture at home, will they not, when exported hinder its consumption and Sale abroad. And if the *English* are not allowed to import, will not the *Dutch* do it so much the more, and thereby hurt proportionably abroad, the Vent and Consumption of our *English* Manufacture.[90]

Davenant had no high opinion of the oriental manufactures, considering that 'Europe draws from thence nothing of solid Use, Materials to supply Luxury, and only perishable Commodities'. But he was equally certain that the government should not end a trade so valuable to the kingdom. English hands should control this business, Davenant asserted, and the profit from supplying the new wants and fashions should go into English coffers.[91] Moreover, he warned against giving too much credence to the imprecations of the 'country party'. Davenant examined the economic foundation of their case and came to the conclusion that,

Some of our Gentry have been for many Years of the Opinion, That the Intire Welfare of *England* depends upon the High Price of Wooll, as thinking thereby to Advance their Rents, but this proceeds from the Narrow Mind, and Short View of such who have all along more regarded the Private Interest of Land, than the Concerns of Trade, which are full as Important, and without which, Land will soon be of little value. . . . And as it [Trade] had never been enough the Study of Our Nobility and Gentry, Who (give me leave to say) for want of Right Knowledge in the General Notion of it, have been frequently Imposed upon, by Particular Merchants, and other Interested Persons, to Enact Laws so much to the Prejudice of Trade in General.[92]

Hoping to forestall any new legislation that might undercut their trade, the Company made efforts to appease critics by shipping wool textiles to India in accordance with the 1693 charter.[93] At best the Company managed to comply with the letter of the charter, but sales of

---

[90] Charles Davenant, *An Essay on the East-India Trade in a Letter* . . . (1696–7), repr. in *Chronicum*, i. 410.   [91] Ibid. 405.
[92] Charles Davenant, *An Essay on the East-India Trade* (1696), 10–11.
[93] *A Short Account of the commencement and of our Commercial Intercourse with the EAST INDIES* . . . n.d., xv.

wool cloth were too small to quiet English merchants or content the weavers.

THE CALICO CAMPAIGN: PROHIBITION OF EAST INDIAN COTTONS

In 1696 the first bill to propose the prohibition of Indian textiles was introduced in the House of Commons, designed to proscribe 'all wrought silks, Bengalls, dyed printed or stained callicoes of the product of India Persia or any place within the charter of the East India Company which shall be imported into this Kingdom', as well as any fabrics of the above mentioned origin 'that are or shall be dyed, printed or stained in this kingdom or elsewhere'.[94] This Act was to be enforced with fines suggested at £100 for traders or retailers in these commodities. Parliament was besieged by wool and silk weavers from London, encouraged to hope for the complete elimination of competitive Indian fabrics following days of testimony by representatives of these trades. Those who testified ascribed all the responsibility for the dwindling of the wool trade to the East India Company, stating that 'the East India trade had so much influence over wool in England that the price thereof rises and falls with the scarcity and glut of the Company's said commodities'.[95] Regardless of these claims, the anticipated legislation was not forthcoming that year, nor in the following year. Both bills faced powerful foes in the House of Lords who hampered efforts to enact this sort of sweeping legislation. Nevertheless, the urgent claims of the wool interests presaged the defeat of the East India Company in its bid to continue as before with unrestricted trade.

The advocates of the wool trade intensified their campaign against Indian imports, citing examples of 'utter Decay of Trade ... that the Master who formerly employed Twenty or Thirty, or more, cannot now employ four, nor find them full work'.[96] A fierce debate ensued. Among the contestants were those who refuted the short-sighted plan to eliminate rival trade-goods as a method of placating local competitors. One suggestion recommended the East India trade be maintained and even encouraged because of the new businesses which sprang from this enterprise. Some losses among the old industries would follow, but the commentator insisted that new employment

[94] *House of Commons Journals*, xi, 496–7, quoted in Thomas, 100.
[95] Thomas, 99–100.   [96] *England's Advocate*, 42–3.

opportunities would subsequently arise. The author of one tract asserted that innovations would provide the competitive edge for English manufactures, overcoming the advantage of low wages enjoyed by the Indian craftsmen. His claims were among the most insightful interpretations of the debate.

> the *East India* Trade, by putting Persons upon Invention, may be the Cause of doing Things with less Labour; and then, tho' Wages should not, the Price of Manufactures might be abated. Arts, and Mills, and Engines, which save the Labour of Hands, are Ways of doing Things with Labour of less Price, tho' the Wages of Men, employed to do them, should not be abated. And the *East India* Trade procures things with less and cheaper Labour, than would be necessary to make the like in *England*; it is therefore very likely to be the Cause of the Invention of Arts and Engines, to save the Labour of Hands in other Manufactures; these are the Effects of Necessity and Emulation . . .[97]

Commentary such as this reflected the critical approach of a minority, prepared to consider what could be achieved through creative effort in a period of invention and innovation.[98] Contemplating the cries of alarm about England's decay, the denunciation of luxury and the excessive spending of the common people, some found nothing to fear but much to applaud in the regular purchases by ordinary folk. Reluctantly, Cary came to similar conclusions after considering the spread of Indian imports: 'if a Manufacture of *Wool* will not please, why not one of *Cotton* . . . no doubt we might in a short time attain to an excellency therein, not only to supply our selves, but also Foreign Markets'.[99] From the discussion and debate it is apparent how profoundly the question of the East India trade shook English society in its assumptions and strictures, both economic and social. The forces massed in support of traditional values and defence of established interests would not be mollified. Their view of England was as a nation where all economic and social activities should sustain proven hierarchies and in no way challenge either the political or economic stability of the nation. Inevitably the multi-faceted menace of the East India trade had to be controlled.

In January 1700 the House of Commons ordered that a bill be

---

[97] *The Advantages of the East India Trade to England, considered*, (1697–8) repr. in *Chronicum*, i. 419–20.

[98] Charles Wilson lists the number of inventions patented in the 1690s at 102, rising from thirty-one in the 1660s. In his estimation the environment of the latter decade 'reflect a period of renewed and feverish interest in industrial invention that was not to be repeated until after 1760, when another great wave of invention and enterprise swept over England'. Wilson, 187–8.

prepared for 'the more effectual employment of the poor and encouraging the Manufactures of England'. The body of the bill was similar in substance to the two previous proposals brought before the House in 1696 and 1697. This time, however, the bill passed through the lower House in a matter of days and then received the sanction of the Lords, who could no longer preserve unchanged the Company's trading rights in England. The first Act for the prohibition of Indian manufactures stated that as of the 29th day of September, 1701,

all wrought silks, Bengalls, and stuffs mixed with silk or herba, of the manufacture of Persia, China or East Indies and all calicos painted, dyed, printed or stained there which are or shall be imported into this kingdom, shall not be worn or otherwise used within this kingdom of England, dominion of Wales, or town of Berwick upon Tweed.[100]

Indian textiles were not banned completely as a result of this first Act; plain unfinished cottons and white materials of every sort could be legally imported. However, the pattern of commercial relations that had developed between England and India during the previous century was checked. This Act was implemented in the hope that new patterns of consumption could be reversed; that cottons could be eliminated from the market-places of England; that clothing, furnishings, and fashion could be returned to the days before commercial expansion, when wool dominated the market without fear of foreign competition. England, however, could not be so easily constrained into a mould of Parliament's making; Indian cottons did not disappear from the kingdom as their opponents had hoped.

The first challenge to this protectionist legislation came from within the manufacturing community of England. Wadsworth and Mann noted that if, 'the first thought of European manufacturers had been prohibition, the second was imitation'.[101] A printing and dyeing industry had grown up in England in the last decades of the seventeenth century, and was given an added boost with the migration of skilled labour from France after 1685. The first mention of a commercial dye-works for cottons can be found in the 1676 petition for a patent of a dyeing process, and Child himself mentioned in 1677 that plain Indian textiles were being coloured in England in imitation of the Indian style. Within a short time the printers in England became so expert that their patterns rivalled those originating in India.

[99] Cary, 7.   [100] 11 & 12 Will. III, c. 10.
[101] Wadsworth and Mann, 118.

Workshops for dyeing and printing sprang up around the London area in response to the continuing demand. The printing industry flourished at Bromley Hall on the Lee in Essex, as well as on the banks of the Wandle and Cray, where bleaching was well established by the late seventeenth century.[102] What began as a small, peripheral occupation advanced rapidly in the conditions after 1700. (See Plate 2, a calico printer.) Defoe recounted with disgust that, 'no sooner were the East India Chintzes and painted calicos prohibited from abroad but some of Britain's unnatural children ... set their arts to work to mimick the more ingenious Indians and to legitimate grievances by making it a manufacture'.[103]

The Commissioners of Trade and Plantations concurred on the evident failure of the Act to curtail the use of foreign cottons when they wrote in 1702 that:

Though it was hoped that this prohibition would discourage the consumption of these goods, we found that allowing calicos unstained to be brought in has occasioned such an increase of the printing and staining calicoes here and the printers and painters have brought that art to such perfection that it is more prejudicial to us than it was before the passing of the Act.[104]

In 1711 English printers computed that they printed one million yards of calico in that year, most of it a coarse variety which cost between 18*d*. and 22*d*. per yard. Even the additional 15 per cent excise tax did not slow the sale of these goods.[105] Wool merchants protested against the extensive sale of these fabrics, insisting that they were

more prejudicial to us than the importation of painted calicoes was before the passing of that act. For whereas then the calicoes painted in India were most used by the richer sort of people whilst the poor continued to wear and use our woollen goods, the calicoes now painted in England are so very cheap and so much the fashion that persons of all qualities and degrees clothe themselves and furnish their houses in a great measure with them.[106]

Far from solving the problems of the wool industry, the 1700 ban on printed textiles stimulated the development of domestic competitors just as determined to secure a position in the home market for their products.

[102] Baker, 46.   [103] Defoe, quoted in Thomas, 121–2.
[104] Report of the House of Lords, MSS, 1702–4, 71, quoted in Thomas, 122.
[105] Wadsworth and Mann, 132–3.
[106] Report of the Commissioners of Trade and Plantations, *House of Lords MSS*, vii, 250, quoted in Wadsworth and Mann, 133.

Manufacturers, as well as printers, turned their hands to imitating Indian wares. In 1702, shortly after the prohibition of printed oriental fabrics came into effect, the East India Company ordered shipments of cotton yarns to be sent to England, 'because the manufacturers have, since the prohibition, attempt the making of Sooseys, Romals, etc., in imitation of Bay goods'. Two years later there was another mention of the industry that developed with the use of Indian yarns. The Company's letter stated that 'weavers here have fell into the use of it in imitating several India manufactures'.[107] Neither the spinners nor the weavers in England possessed the skill to produce comparable fabrics in pure cotton, so the British mixed wool or linen with cotton in its manufacture. The weavers in Spitalfields originated these early copies, for they were skilled in reproducing expensive foreign fabrics. But they did not retain this trade. Lancashire already specialized in linen and fustian weaving and several sorts of these fabrics were mentioned in the 1696 directory of textiles *The Merchant's Warehouse laid open*. Manufacturers in the north-west attempted to make various kinds of textiles that would not only compete successfully against imported European linens, but would also make inroads into the market formerly filled by East Indies goods. In 1712 the equalization of taxation between the European-made checked and striped linen and the cotton-linen cloth made in Lancashire provided an incentive to compete for a greater part of the home market. Six years later the Lancashire industry was flourishing and competitive European fabrics were all but ousted from the English market.[108]

The nascent cotton trade in Lancashire could not yet match India for variety or quality. None the less, the early English cottons shared crucial features with the Indian materials on which they were modelled. Variety in price, patterning, and colour were key characteristics of the new consumer textiles, beginning with the New Draperies and carrying through to the East Indian and later those fabrics manufactured in England. Durability was no longer the most desirable attribute; originality and cost predominated as the most decisive features. English-made products could not completely replicate those from India; but increasingly, proficiency was gained in weaving and printing goods that appealed to the public for exactly the same reasons that had made calicoes and chintzes the rage. The wool interests had not extinguished this craze. On the contrary, ambitious and inventive

---

[107] East India Company's Letter Book, x, fo. 538, xi, fo. 414, quoted in Wadsworth and Mann, 126.   [108] Wadsworth and Mann, 126–7.

craftsmen and tradesmen sought to duplicate the success of Indian textiles with a local product.

THE LAST CALICO CAMPAIGN

The wool trade had always been sensitive to any disruptions in its affairs, nationally and internationally. Characteristically, wars produced a cyclical surge then slump, as orders for the armed forces were followed by closing of overseas markets. Reduced orders led to a slowing of employment back in England. The War of the Spanish Succession (1701–13) followed the pattern of previous European wars, whereby an initial flurry of orders and a sharp rise in demand for fabrics encouraged merchants and manufacturers to over-produce. Defoe disputed the cry of a decay in trade, scolding the clothiers for their intemperance.

Upon some sudden Accident in Trade here comes a great unusual Demand for Goods, the Merchants from Abroad have sudden and unusual Commissions, and the Call for Goods . . . encreases . . . The Country Manufacturer looks out sharp, hires more Looms, gets more Spinners, gives more Wages, and animated by the advanc'd Price . . . gluts the Market with Goods. The Accident of Trade . . . being over, those Demands are also over, and the Trade returns to its usual Channel; but the Manufacturer in the Country, who has run out to an unusual Excess in his Business . . . having not stopt his Hand as his Orders stopt, falls into the Mire; his Goods lye on Hand . . . then they cry our Trade is decay'd, the Manufacturers are lost, Foreigners encroach upon us, the Poor are starv'd and the like.[109]

The response of the wool trade to the War of the Spanish Succession was exactly as Defoe described it, leading to a wave of unemployment and a cry against the 'foreigners' usurping English trade. Wool manufacturers continued to stagnate, suffering the disruption and dislocation caused by war, exacerbated by the development of rival wool industries in former markets. The powerful political connections of the wool and London-based silk interests focused their energies on ridding the market of Indian cottons.[110] The campaign against the sale

---

[109] Daniel Defoe, *A Plan of the English Commerce*, 2nd edn. (1730, repr. 1967), 257–8.

[110] The silk industry in England also claimed to be a national trade deserving protection from East India imports, and in fact experienced very direct competition with some sorts of oriental fabrics. See Natalie Rothstein, 'The Calico Campaign of 1719–21', *East London Papers* (July 1969), where she focuses on the campaign of the London silk industry.

and use of cottons in England did not continue unabated for the twenty years between the first Act in 1700 and the second in 1721. Rather, the agitation rose and fell with the health of the wool trade; each reversal in sales bringing a renewed clamour for the elimination of the heathen fabrics from the kingdom.

In 1718 war once more engaged Britain's attention in Europe. The campaign against Spain received formal acknowledgement with a declaration of war in December of that year. Shortly thereafter another declaration of war was announced, but this time within the kingdom. On 10 June 1719, a man named John Humphreys led hundreds of weavers from the silk-weaving district of Spitalfields in a general assault on all wearers and sellers of the hated calicoes. The battle-lines were drawn; the prize to be won, the home market. On 11 June four Companies of militia were raised to put down the first wave of disturbances. Major Hardwick later testified that:

on the 12th they [the militia] gathered together again and he having advice that ... [the weavers] intended to pull down a House in Brown's Lane, wherein were some Women in Callicoe Gowns, marcht with his Company to prevent it; that he order'd his Lieutenant to enter the Lane with part of his Company at one end, while he and the rest marcht in at the other, which they did accordingly; that he took Bains and Picket clad in Callicoe to the Waste, whom he took to be the Ringleaders ... they cry'd *down with the Callicoes* ...[111]

The outbreak of this disturbance on 10 June caused general dismay, as the date was the anniversary of the insurrection in support of James Stuart several years earlier. It seemed possible that the complaints of the weavers could turn into a more general uprising, for the weavers were soon 'joyn'd by many Idle Fellows who had no other View *but* Plunder'.[112] John Humphreys was also accused of previously 'Speaking Seditious Words against his Majesty' in an incident in 1717, when he reportedly denied that King George was the rightful heir to the throne, drinking a toast to the Pretender. Subsequently, witnesses were found who tied Humphreys in with the unrest.

Benjamin Horn deposed, that as he passed through Wide-Gate-Alley, seeing a Woman go before him drest in a Callicoe Gown, said, it was a pitty they should be wore; whereupon John Humphreys came up to him and ask'd him to drink, and told him that he had spent Five or Six Guineas that Day to encourage the

---

[111] *Proceedings on the King's Commission of the Peace ... for the City of London and ... for the County of Middlesex; held at ... the Old Bailey* (hereafter referred to as the *Old Bailey Records*), July 1719, 7.   [112] *The Flying-Post*, 11–13 June 1719.

Mob, and that he hoped to have Ten for them the next Day; that they should be reinforc'd; the Norwich Weavers were coming up, and then the Business would be done.[113]

Enraged weavers spread out from the city centre, attacking printing shops and any other sites associated with printed cottons. Neither were ordinary people safe from attack if they had the good taste, but political misfortune, to be wearing patterned or printed clothing. On 13 June the *Weekly Journal* recounted the ongoing violence whereby

> the Weavers ... committed several Disorders and Outrages on the Bodies of Persons wearing Callicoes, and printed Linnen, and burn'd all such sort of goods as they could get out of the Shops where they were to be sold. They likewise design'd to have ransack'd the House of M. Movillion, a Callico Printer near Windsor, but the Horse Guard came up time enough to prevent it.

The writer concluded that the 'Damage done by those Rioters in several Parts of this City, appears to be very considerable; for they spar'd none, even those of best Fashion who were Callico, either tearing or spoiling them with Aqua-Fortis, Ink, etc'.[114] The effects on the victims of these attacks can be imagined. Elizabeth Price was in the parish of St Leonard Shoreditch looking for lodgings. She recounted that as she went to look at the house, 'some People sitting at their Doors, took up her Riding Hood, and seeing her Gown, cry'd out Callicoe, Callicoe; Weavers, Weavers. Whereupon a great Number came down and tore her Gown off all but the Sleeves, her Pocket, the head of her Riding Hood, and abus'd her very much'. Along with her gown, Elizabeth lost the guinea that was in her pocket; neither was ever recovered.[115]

During the following weeks public disturbances multiplied, and assaults on citizens and property continued. A company of Guards was sent to contain and quell the rioting weavers and their partisans, arresting several and finally firing into the crowd, killing and injuring several of their number. But even this level of repression could not ensure an end to civil strife. Rumours circulated in London that the weavers would soon be joined by 'a large body' of Norwich weavers, come to aid in the campaign to rid the country of calicoes. That this rumour proved to be false was some consolation to the Lord Mayor

---

[113] *Old Bailey Records*, July 1719, 7.
[114] *Weekly Journal*, 13 June 1719.
[115] *Old Bailey Records*, July 1719, 7.

and his advisors; however, riots flared in various parts of London and in other manufacturing centres. At Bunhill Fields four men wearing stolen pieces of calcio tied around their waists were arrested. When these men were seized the crowd surged in about them, blocking the guards' retreat, and from the centre of the crowd cries of 'no Hanoverian King' were heard. Outbursts such as these warned of a dangerous climate of disaffection. Measures had to be taken to placate and control the inflamed artisans.[116]

In early July the Company of Weavers of London appealed to their fellow journeymen and apprentices to refrain from violence. Reports from Norwich indicated that the riots and disorders there had been contained, if not eliminated, by the middle of June. In fact, the reimposition of order there was far from complete, for it was said that the weavers, 'will not it seems yet let any Body pass the Streets in Callicoe without Mobbing of them; and as the Complaints of the Want of Trade is still very great there, 'tis said the Magistrates keep strict Watch to prevent the like Rising for the future'.[117] Spokesmen for the wool trade delivered homilies deploring the violence, but explained the behaviour of the weavers as being a desperate effort to preserve their trade. Critics pointed out that Parliament was not sitting and so could not judge the merits of their case, drawing attention to the 'Treasonable Practices that were prov'd upon [Humphreys]' following his arrest and trial at the Old Bailey.[118] Although Humphreys was not himself a weaver he had by this time gained the acclaim of the body whose cause he espoused. On 1 August, he and three others convicted of riotous behaviour were placed in the pillory at Spitalfield Market, surrounded by a mass of silent weavers, watched in turn by a detachment of militia. The eruption of this gathering resulted in further prolonged disturbances; a sympathetic weekly newspaper chronicled the sequence of events.

The Weavers look'd on [the three men], but committed no Irregularities, till some Women had the Folly to appear in a Hackney Coach dress'd up in Callico, at which the poor Men being too much moved for their small Stocks of Patience to govern, they offered Rudeness enough to their Callico and tore it pretty much, some say stript it off their Backs, and sent them home to dress over again.[119]

[116] *Weekly Journal*, 20 June 1719; *The Flying-Post*, 25–7 June 1719.
[117] *The White-hall Evening Post*, 7–8 July 1719; *Weekly Journal*, 18 July 1719.
[118] *A Further Examination of the Weavers Pretences* (1719), 4.
[119] *Thursday's Journal*, 6 Aug. 1719.

Over the following weeks Whitechapel, Stepney Fields, Moorfields, and the area by London Bridge were convulsed by individual and group violence, usually occasioned by the sight of a woman wearing a printed gown.[120] Women so attired walked in fear of being mobbed and having the clothes torn from their backs. Reports also described how male friends and husbands vigorously defended the women from these onslaughts, precipitating general mêlées as weavers, wives, and husbands fought; the women in defence of their clothing, their modesty, and their right to choose; the consorts in defence of their women; and the weavers for the right to dictate common fashions, one way or another. Despite such rough and ready arbitration there was considerable public support for the demands of wool and silk weavers, though the printed tracts and pamphlets advised restraint, so that the just cause of the weavers could be judged on its own merits. But there is little doubt that the violence was itself effective in attracting attention and generating concern both about public order and the root cause of the weavers' complaints.

The ire of the wool industry was directed not only against imported calicoes, but against all the printed fabrics worn in England, whether English-made or imported. The common characteristics of all these materials was their light weight, colourful prints, and washability, characteristics that were not uniformly available in wool textiles. One author cautioned his readers from assuming that a change in tastes could be imposed on the English population, noting that

> all those who wear Callicoe or Linen now, wou'd not wear Woollen Stuffs if there was no such thing as Printed Callicoe or Linen, but *Dutch* or *Hambro'* Strip'd and Chequer'd Linens, and other things of that kind, and for the same Reason that they now wear printed Callicoe or Linen ... because nothing else washes near so well.[121]

In spite of any evidence to the contrary, the pamphleteers in the employ of the wool and silk interests insisted that tastes could be altered through a legal bar on competitive goods:

> The very Weavers and Sellers of Callico will acknowledge, that all the mean People, the Maid Servant, and indifferently poor Persons, who would otherwise cloath themselves, and were usually cloathed in thin Womens Stuffs made at *Norwich* or *London*, in Cantaloons and Crapes, etc. are now cloathed in

---

[120] *The Weekly Medley*, 8–15 Aug. 1719; *Weekly Journal*, 8 August, 1719.
[121] *A Further Examination of the Weavers Pretences* ..., 20.

Callicoe, or printed Linen; moved to it as well for the Cheapness, as the Lightness of the Cloth, and the Gaiety of the Colours. The Children universally, whose Frocks and Coats were all either made of Tammies worked at *Coventry*, or of striped thin Stuffs made at *Spittlefields*, appear now in printed Callicoe, or printed Linen; let any one but cast their Eyes among the meaner Sort playing in the Street, or of the better Sort at Boarding School.[122]

The English weavers sought to rally public sentiment behind their fight against the printed fabrics and to shame or browbeat women into giving up these articles.[123] In spite of pleas for restraint, protesting weavers carried on their campaign, breaking into shops and houses in search of any article of printed textiles and parading through the streets shouting 'King George forever' to counter claims of sedition. Demonstrations fell off as winter approached and the market for woollens improved. By then, however, the government was convinced that some action had to be taken. Concerns about public disorder spurred government activity. Other arguments and claims galvanized the landowners in Parliament, who flinched at the political and financial cost of a further decline in the wool trade, believing, as did one editorialist, that:

when the weaving Trade is dead; all other Trades are dead, I mean all English Trade, and when the Poor wants imployment 'tis then their Parish suffers; for every one of these Poor people so imploy'd; the Wages they have is laid out again with their Neighbours for such things as they have Occasion for, so that what they get by their Labour serves not only to Supply their own Wants but every one hath an Advantage thereby. Sir, 'tis a pity that the Wool; which is the only Jewel this Nation affords should be so much dispised, it being that only which have and must support this Nation. . . . It is not the multitude of Weavers that cause the Trade to be dead, no but the multitude of foreign

---

[122] *The just Complaints of the poor Weavers truly represented*, repr. in *Chronicum*, ii, 195.

[123] Taking the persona of an aggrieved woman, Defoe wrote an article that attempted to defend the weavers' ideals, if not their tactics. But he raised an interesting argument circulating during that time, without offering a cogent solution to the question raised. Did ordinary citizens have the right to select clothing according to personal preference and the prevailing fashions? Defoe sympathized with the 'fellow' women, who had been, 'abused, frighted, stript, our Clothes torn off our Backs every day by Rabbles,—under the pretence of not wearing such Clothes as the Weavers please to have us wear'. This character went on to argue that, 'We always thought, and have been told by our Grandfathers, that English people enjoyed their lawful Liberties above all the Nations in the World . . . Never tell us of National Liberties! If our Sex has not a Share in the Liberties, how can they be National?' Defoe, *Weekly Journal*, repr. in William Lee (ed.), *Daniel Defoe and His Life and Recently Discovered Writings . . .* , *1716–1729* (1869, repr. 1969), ii, 140.

Callicoes and printed Linnen that is worn among us; and were they once removed, you would certainly find that there would be Imployment enough for the Weavers.[124]

Among the hundreds of pamphlets, petitions, and tracts which demanded a ban on printed textiles there was no apparent realization that the English market was changed and would continue to demand articles consistent with popular style. (See Plate 3 for an example of such a tract.) There was no ostensible appreciation of the power of consumer preference, nor was there an acknowledgement of the benefits of diversifying manufacturing, lifting the burden of British industry from the back of the sheep on which it had rested for so many generations. The calico campaign was an exercise in conservatism. What no one had yet discussed was the way in which the general population could be forced to accede to the sumptuary restraints demanded by the wool and silk industries. What mechanism did the state have to restructure popular fashions, to reconstitute consumer tastes? Only the crudest sort of legislative prohibition was at the disposal of the English Parliament as they considered ways and means of alleviating the malaise of artisans, traders, merchants, and pastoralists.

The campaign was kept charged during the winter months with a heated debate between the forces of prohibition and the faction supporting unrestricted commercial access to the British market. One petition to the ladies of England from 'Dorothy Distaff' begged that an end be made to 'the Wearing of these Callicoe Trumpery', hoping simplistically that 'As the general Wearing of Callicoes is the Complaint, the general Leaving them off will be the Cure'. Calicoes were castigated as being a 'tawdry, Pie-spotted, flabby, ragged, low-priz'd Thing . . . made . . . by a Parcel of *Heathens* and *Pagans*, that worship the Devil, and work for a Half-penny a Day'.[125] Not all the petitioners showed such a flamboyant turn of phrase. Those writing on behalf of the East India trade, linen merchants, calico printers, English and Scottish linen producers, pointed out the justice of their claim to a share of the rich home market. In particular, British manufacturers of linens and cotton-linens disputed the right of silk manufacturers to describe their trade as national. Silk came from no part of the British Empire. In contrast, cotton used in fustians and English-made calicoes

[124] *Weekly Journal*, 22 Aug. 1719.
[125] *The Female Manufacturers Complaints: Being the Humble Petition of Dorothy Distaff*. . . (1720), 10, 13, 16.

came from British plantations, the flax was grown on British soil, and the whole was made entirely within the kingdom.[126] Charges and counter-charges levied before the members of the House demanded some response.

Initially a bill was introduced in the Commons banning all printed cottons and linens. Even in the face of the massed fury of the weavers, who ringed the House and made it echo with their clamour and groans, the Lords were not prepared to accept so precipitous an action. They postponed their decision. In the next sitting of Parliament the claims of English cotton and linen manufacturers were placed before the House. Petitioners insisted that 'Encouraging and Carrying on a Manufacture of Callicoe here, will make a very great Addition to the Trade of the Nation . . . imploying many Thousands of the Poor'.[127] This contention found few supporters; as yet there were no powerful advocates to plead the case of a domestic cotton industry. In the anxious economic climate following the South Sea Bubble, Parliament was disposed to act to bolster the wool trade.

In 1721 cotton textiles of almost every description were banned from England.[128] Both the use and sale of painted, printed, flowered, checked, and most of the other vibrant Indian textiles were no longer legal either in dress or furnishings. Information brought to Justices of the Peace by a sworn witness would result in fines of £5, that would go to the informer; the moneys to be raised through the sale of the prohibited goods. Mercers and drapers discovered to be selling calicoes would suffer a £20 fine, as too would buyers of these goods.[129] The champion of wool interests, the *Manufacturer*, extravagantly predicted that from 'this day we may date the resurrection from the dead, as well of our foreign declining commerce as of our home

---

[126] *The Weavers Pretences Examined* . . .; J. Asgill, *A Brief Answer to a Brief State of the Question, Between the Printed and Painted Callicoes, and the Woollen and Silk Manufacturers* (1720); *The Case of the Convention of the Royal Boroughs of Scotland, in relation to the Linnen-Manufactory of that Country* (1720); *The Case of the Printed Linnens of North Britain* (1720?).

[127] *Reasons for adding a Clause to the Bill for Preserving and Encouraging the Woollen and Silk Manufacturers, etc. to except Callicoe Manufactured in Great Britain, out of the intended Prohibition* . . . (1720?).

[128] 7 Geo. I, c. 7. This ban would not be lifted by the House of Commons until 1774. By that time the British cotton industry was in its ascendency. Richard Arkwright petitioned Parliament to remove the restraint on British manufacturers. It was easy to convince the legislators of the need to free the cotton industry from restriction no longer in the interest of British industry. See Wadsworth and Mann, 485.

[129] Malachy Postlethwayt, *The Universal Dictionary of Trade and Commerce*, 4th edn. (1774), vol. i.

manufacture', and applauding the wisdom of the legislators expressed 'the felicity of being born Britons'.[130] Others, less sanguine about the Act, had already expressed reservations about its efficacy and wisdom. One in particular noted with irony that, 'many Nations Sumptuary Laws have been made to restrain the Peoples Expence, but it seems strange to oblige Persons to a greater Expence in their Apparel then they are willing, or can afford'.[131]

In practice, Indian calicoes were not swept from England. (See Plates 4 and 5 for an example of a printed Indian gown, *c.*1740, the like of which continued to be in demand in Britain.) Nor had the wool and silk industries eradicated competitive printed textiles from the home market. Parliament had taken the most stringent measures at its disposal in an effort to enforce conformity among English consumers, hoping to compel a change in taste and a return to older consumer habits. Far from abolishing the threat of competition, the 1721 Act channelled the stimulus of demand towards domestic manufactures. Printed fabrics made by English craftsmen were sheltered by the Act from the superior manufactures of the Indies. The manufacturers of cotton-linen fabrics were willing and able to increase production to meet a desire for goods that continued unabated, for 'the Humour of the People [runs] so much upon wearing painted or printed Callicoes and Linnens'.[132] This 'humour', however distasteful to the wool trade and its allies, could not be legislated out of existence. Parliament passed two Acts that were intended to preserve ancient privilege founded on wool. The first skirmish was won by the old alliance; but the campaign would not end here. The power of fashion and the attractions of new consumer goods were too powerful to be so easily eradicated and continued to affect the structure and development of a domestic industry over the course of the eighteenth century.

[130] *Manufacturer*, 1720, quoted in Thomas, 159.
[131] *Some CONSIDERATIONS relating to the desired Prohibition of wearing printed CALLICOES, by a Person wholly disinterested otherwise, then for the good of his country* (1720).   [132] *Weekly Journal*, 27 June 1719, repr. in Lee, *Defoe*, 137.

2
DOMESTIC DEMAND IN THE EIGHTEENTH CENTURY: THE BRITISH MARKET FOR COTTON TEXTILES

CHARACTERISTICS OF THE DOMESTIC MARKET

THE British market supported and nurtured the cotton industry throughout the eighteenth century. Thus, an assessment of the domestic market is key when considering the process of development of this industry. Paradoxically, the home market has been recognized as vital to the nascent cotton industry, although little time has been spent on its study when compared to the export markets.[1] Nevertheless, it was the former environment which both protected and succoured the British cotton industry.

Discussion of domestic demand must begin with an appraisal of eighteenth-century Britain, focusing on fundamental factors such as the population structure, movement of wages, plus other elements which influenced levels of demand. The absence of abundant, comprehensive records compels historians of the eighteenth century to acknowledge the ever-present limitations in all quantitative economic analyses. Because of these difficulties, efforts made to discern the structure of that society must include the speculative as well as the proven. On the question of population, however, much less imagination is now required than in previous decades. *The Population History of England* by E. A. Wrigley and R. S. Schofield reveals both the population trends and principal demographic characteristics throughout this period. According to their calculations, Gregory King's estimate of five-and-a-half million people closely approximated the numbers living in England at the beginning of the century, with only a slight over-estimation.[2] Their estimates suggest a slow but regular rise

---

[1] M. M. Edwards, for example, noted that the British market was a 'stable base' for the cotton industry, 'less volatile, subject to fewer crises, and considerably easier to serve' than foreign markets. *The Growth of the British Cotton Trade, 1780–1815* (1967), 27.

[2] Gregory King, *Natural and Political Observations and Conclusions upon the State and Conditions of England* (1696), in G. E. Barnett (ed.), *Two Tracts* (1939), 18.

in the country's population, particularly after the 1740s. This increase was attributed to a rising birth rate which reshaped the structure of society, leading to the highest proportion of young dependents to adults yet seen. Table 2.1 summarizes the movement of population in England over the century, using the data distilled by Wrigley and Schofield from the samples of parish registers across England.[3] Such a continuous sustained advance in population was unprecedented in modern history, transforming the nation, adding larger contingents of babies, children, and adolescents to the society with each decade. Parallel with the rising curve of population was the demand for food, clothing, housing, the necessities and niceties of life, powering the economy, fuelling demand for agricultural as well as industrial commodities.

Growing pressure on wages was one of the most crucial of the repercussions felt as a result of the rise in population. Up until the 1750s real wages rose steadily as a by-product of expanding agriculture and vigorous commerce in conjunction with a slowly ascending birthrate. In the first half of the eighteenth century bumper harvests brought greater prosperity and cheaper food prices within reach of all. Living standards improved generally, prompting a commensurate elevation of expectations which carried through into the second half of the century, where substantially new conditions prevailed. For contemporaries the years of prosperity brought more and more varieties of affordable consumer goods and a greater degree of comfort than had been enjoyed within living memory. Under such conditions a general optimism infected the population: the atmosphere was one of vitality and hope for continued plenty. No figures better reflect this

TABLE 2.1. *Estimates of population for England, 1701–1801* (millions)

1701	5.058	1761	6.147
1711	5.230	1771	6.448
1721	5.350	1781	7.042
1731	5.263	1791	7.740
1741	5.576	1801	8.664
1751	5.772		

[3] E. A. Wrigley and R. S. Schofield, *The Population History of England, 1591–1871* (1981), App. 5, 577.

national conviction than the age of first marriages. Marriages could be entered into only when a couple had accumulated sufficient resources to support an independent household free from parental assistance. Times of hardship, famine, unemployment always witnessed a drop in marriages; while in times of abundance marriages occurred more frequently and at an earlier age.[4] Common expectations determined when and at what age marriages were contracted, providing a barometer of national expectation that can be gained from no other source. Over the course of the eighteenth century not only was the age of first marriage falling, but a larger proportion of the population entered into marriage, and both factors propelled the birth-rate upwards.[5] The men and women of England looked to the future with hope, and acted on anticipated continuing prosperity. For as Wrigley and Schofield aptly observed: 'Precisely because marriages were for a long time it is not simply short-term prospects that affect marriage decisions but an appreciation of the probable course of events over half a lifetime'.[6]

Growth of population exerted pressure on real wages. There had always existed a delicate balance, typical of an agrarian-based society, that allowed the level of real wages to remain stable or even rise slightly, as long as the population-increase did not exceed 0.5 per cent annually. Once that rate of increase was exceeded a corresponding slide in real wages ensued.[7] This cycle of action and reaction came into play shortly after the middle of the century; nevertheless, the trend to earlier marriages persisted. Table 2.2 charts the progressive drop in the age of first marriage.[8]

The decline in the age of first marriage, from 1750 to 1800, of over one year for men and nearly one-and-a-half years for women argues

TABLE 2.2. *Age of first marriage through the eighteenth century*

	Male	Female
1700–24	27.3	25.8
1725–49	26.9	25.6
1750–74	26.2	24.7
1775–99	25.9	24.1

[4] Wrigley and Schofield, 421.
[5] Ibid. 267, Table 7.29.
[6] Ibid. 421.
[7] Ibid.
[8] Ibid. 424.

strongly that expectations remained high and that people acted on the belief that they could sustain new households. Successive generations married at a younger age and had more children.[9] Moreover, the broadest popular expectations were fulfilled. For if, as T. S. Ashton asserted, 'the central problem of the age was how to feed and clothe and employ generations of children out-numbering by far those of any earlier time', then Britain was ultimately successful.[10]

The pressures of population halted the wage-rise early in the second half of the century. Shortly thereafter, a decline in real wages can be observed. Wrigley and Schofield have drawn attention to a unique economic feature: for in spite of the pressures on real wages exerted by the expanding labour supply, around the last decade of the century the typical response of an agrarian society did not materialize. There was no crisis; there was no catastrophic fall in wages. The correlation between real wages and population, characteristic of a land-based economy, no longer applied. The impending crisis anticipated by Malthus was averted through the power of the newly unleashed industrial might. Wrigley and Schofield describe this unique departure:

Until the radical change at the end of the eighteenth century rapid population growth, at a rate of 0.5 per cent per annum or more, entailed falling real wages, whereas periods of slight population growth or population decline were associated with rising real wages. Towards the end of the eighteenth century, however, although population was growing at more than 1 per cent per annum, and significantly faster than at any earlier period, real wages, which had been falling sharply, recovered and began to rise also, reaching a rate of growth higher than any previously experienced in spite of the rapidity of population growth. *For the first time since land was fully settled rising numbers proved consonant with rising wages. The meaning of the industrial revolution is visible in the figure.*[11]

The evidence of the declining age of first marriage provides impressive, though dumb, testimony from the men and women of the age. In their eyes the economy appeared vigorous; diversification of employment must have suggested an economy sufficiently expansive to justify early marriage. Wrigley and Schofield discerned a major social phenomenon reflective of the general economic status of the nation,

---

[9] W. A. Cole speculated that 'rising expectations may have been engendered by the increasing real wages of the first few decades of the century'. Expectations might continue high, even when wage-rates did not. W. A. Cole, 'Factors in demand, 1700-80', in Roderick Floud and Donald McCloskey (eds.), *The Economic History of Britain Since 1700* (1981), i, 58.

[10] T. S. Ashton, *The Industrial Revolution* (1948, repr. 1975), 161.

[11] Wrigley and Schofield, 409 (my emphasis).

contradicting some of the pessimistic hypotheses on the movement of real wages throughout the period. Furthermore, the movement of wages did not constitute the only influence on the earning capacity of the nation's families, and may not have been the most dynamic feature of this period. One can only conjecture as to all the possible local influences individual trades would have exerted. Indeed, average national indicators are known to hide the highs and lows of what was essentially an aggregate of regional economies.[12] Moreover, it must be noted that when speculating on the movement and impact of real wage-rates, 'it is ultimately impossible to accommodate the complexity of historical experience within the strait-jacket of any single economic model'.[13] The scarce and random documentation that survives can only suggest trends. Thus an appraisal of potential demand within the economy must recognize the myriad personal, local, and regional influences that affected family income and the state of the home market during this period.

Peter Lindert and Jeffrey Williamson have presented a substantially revised wage-scale based on the original work done by Phelps Brown and Hopkins.[14] In the words of Lindert and Williamson, their work 'mines an expanded data base and emerges with a far clearer picture of worker's fortunes from 1750'.[15] They have computed the most detailed chart to date of estimated earnings for a total of eighteen sectors of employed adult males. The wage-scales are augmented from many sectors other than the Phelps Brown and Hopkins building trades, including the estimated annual earnings of those employed in

---

[12] Intensive investigation of rural economies on the local level have produced dramatic indications of the complexity and diversity of these economies. See the recent volumes in *The Agrarian History of England and Wales*, vol. v, p. 1, and in particular the introduction by Joan Thirsk. In this she briefly describes the range of agricultural and economic environments at work in England between 1640 and 1750, many of which developed in association with towns or expanding industries. Regional economies determined wage-rates, local demand, and the overall prosperity of the area. Certainly national trends were also at work—population growth, internal and external trade—but more recent research suggests local characteristics must be integrated into any generalized assessment of the eighteenth-century economy.

[13] Cole, 63.

[14] The Phelps Brown and Hopkins wage-scale was used by Wrigley and Schofield in their recent volume, though with some reservations, due to the narrow range of trades used as a database and the problems that resulted when these figures were extrapolated with the existing cost-of-living statistics.

[15] Peter H. Lindert and Jeffrey G. Williamson, 'English Worker's Living Standards during the Industrial Revolution: A New Look', *Economic History Review*, 2nd series, 36: 1 (1983), 1.

service industries—a vital expanding area in the eighteenth century hitherto unrepresented in any wage-scale (see Table 2.3). Beginning in 1755, estimates of annual earnings for the eighteen occupational groups are provided. These yield the widest range of estimated wage-rates for the period and suggest interesting patterns for expanding consumption. Trends in real earnings among adult males were found to have suffered post-1750 from the effects of population increase. However, they observed for the first time a variable influence among the representative working groups. All levels experienced the effects of declining real wages post-1750, although farm and non-farm labourers began to catch up with the more highly paid skilled labourers before 1800. The middle group of the eighteen occupational sections considered, along with those in professional or white-collar occupations, suffered comparatively little from declining real wages, while artisans experienced the biggest drop.[16] All of these generalizations take into account the varying impact of the movement of real wages, for a comparable percentage drop in wages would have more stringent effects on the unskilled labourer earning £20 per year than on a clerk earning £50. Lindert and Williamson's analysis of real wage-rates offers a challenge to the broadly pessimistic scenario of living standards during the last decades of the century, and, by inference, the estimated levels of demand.

Britain was a patchwork of wage-rates. The singularity of local conditions must be kept in mind at all times, as historians grapple with the question of patterns of income and collective demand. Within any one region the range of work and remuneration available provided ascending scales of earnings. London wages, for example, were one-third above those in next-door Kent.[17] The new and burgeoning industries also affected the lives of many, directly and indirectly. This took is a decisive consideration in the study of the home market for British-made cottons, for a general impoverishment of those employed by the new industries would have eliminated from the roll of consumers those likely to be in a position to acquire some of the fruits of these trades. Understandably, internal migration favoured areas where new skills or old could command comparatively better rewards. The north and midlands generally offered higher wages than southern Britain. It has long been noted that wages were influential in drawing

---

[16] Peter H. Lindert and Jeffrey G. Williamson, 7, 13.
[17] Gilboy, quoted in P. K. O'Brien and S. L. Engerman, 'Changes in income and distribution', *The Economic History of Britain since 1700*, i, 167–8.

TABLE 2.3 *Estimates of nominal annual earnings for eighteen occupations, 1755–1851: adult males England and Wales (in current £'s)*

Occupation	1755	1781	1797	1805	1810	1815	1819	1827	1835	1851
(1L) farm labourers	17.18	21.09	30.03	40.40	42.04	40.04	39.05	31.04	30.03	29.04
(2L) non-farm common labour	20.75	23.13	25.09	36.87	43.94	43.94	41.74	43.65	39.29	44.83
(3L) messengers & porters	33.99	33.54	57.66	69.43	76.01	80.69	81.35	84.39	87.20	88.88
(4L) other government low-wage	28.62	46.02	46.77	52.48	57.17	60.22	60.60	59.01	58.70	66.45
(5L) police & guards	25.76	48.08	47.04	51.26	67.89	69.34	69.18	62.95	63.33	53.62
(6L) colliers	22.94	24.37	47.79	64.99	63.22	57.82	50.37	54.61	56.41	55.44
(1H) government high-wage	78.91	104.55	133.73	151.09	176.86	195.16	219.25	222.95	270.42	234.87
(2H) shipbuilding trades	38.82	45.26	51.71	51.32	55.25	59.20	57.23	62.22	62.74	64.12
(3H) engineering trades	43.60	50.83	58.08	75.88	88.23	94.91	92.71	80.69	77.26	84.05
(4H) building trades	30.51	35.57	40.64	55.30	66.35	66.35	63.02	66.35	59.72	66.35
(5H) cotton spinners	35.96	41.93	47.90	65.18	78.21	67.60	67.60	58.50	64.56	58.64
(6H) printing trades	46.34	54.03	66.61	71.11	79.22	79.22	71.14	70.23	70.23	74.72
(7H) clergy	91.90	182.65	238.50	266.42	283.89	272.53	266.55	254.60	258.76	267.09
(8H) solicitors and barristers	231.00	242.67	165.00	340.00	447.50	447.50	447.50	522.50	1166.67	1837.50
(9H) clerks	63.62	101.57	135.26	150.44	178.11	200.79	229.64	240.29	269.11	235.81
(10H) surgeons & doctors	62.02	88.35	174.95	217.60	217.60	217.60	217.60	175.20	200.92	200.92
(11H) schoolmasters	15.97	16.53	43.21	43.21	51.10	51.10	69.35	69.35	81.89	81.11
(12H) engineers & surveyors	137.51	170.00	190.00	291.43	305.00	337.50	326.43	265.71	398.89	479.00

Sources and Notes: From Peter H. Lindert and Jeffrey G. Williamson, 'English Workers' Living Standards during the Industrial Revolution: A New Look', *Economic History Review*, 2nd series, 36: 1 (1983), 4. Some of these occupations need no elaboration. Those that do are explained as follows: (4L)—watchmen, guards, porters, messengers, Post Office letter carriers, janitors; (1H)—clerks, Post Office sorters, warehousemen, tax collectors, tax surveyors, solicitors, clergymen, surgeons, medical officers, architects, engineers; (2H)—shipwrights; (3H)—fitters, turners, iron-moulders; (4H)—bricklayers, masons, carpenters, plasterers; (6H)—compositors.

local labour from agriculture to the factory. Local landowners and other employers were eventually obliged to offer sufficient financial inducement to retain their workers; in so doing they helped create a high-wage enclave throughout the industrial regions. The absence of any migration away from the developing industrial provinces suggests that those who laboured in mills, workrooms, and potteries found some compensation in the wages paid them, sufficient to endure the new discipline and the often deplorable conditions of work.

In any discussion of incomes it must also be noted that money wages commonly comprised only part of the payment for work undertaken in early industrial Britain. Among the wage-earners in mills and factories one found labourers whose incomes were composed entirely of cash wages.[18] Their numbers would be few, matched against the millions employed in non-industrial occupations. Throughout the eighteenth century and beyond, traditional perquisites accompanied most sorts of employment.[19] The domestic servant received used clothing from master or mistress, while those employed in kitchen, garden, stable, or field expected other sorts of goods to supplement their wage. Perquisites were an expected and accepted method of payment, though only infrequently recorded. The constraints imposed by declining real wages over most of the last decades of the century would have been buffered to some extent by these additions. Whether a cottager's garden, or second-hand linen or clothing, such payments in kind provided the worker with goods either to sell, use, or barter, to improve the position of the family and add purchasing power to the family unit.[20] Such additions to income have not been commonly computed by the application of economic models. These payments are

---

[18] Here too this was not always the case, particularly in the early industrial period when it was sometimes difficult to attract workers to rural areas where the water-driven jenny mills were located. Samuel Oldknow certainly suffered from this problem and offered cheap housing to attract workers. Thomas Evans was more forthcoming, advertising garden plots and promising a milk-cow for every family in his employ. S. D. Chapman, *The Early Factory Masters* (1967), 157–9.

[19] In George Ewart Evans, *Ask the Fellows that cuts the Hay* (1956), the oral histories of Suffolk labourers confirm the importance of perquisites and non-cash payments well into the twentieth century. Articles such as portions of mutton, fuel of various sorts, special egg-money, and the like were indispensable additions to family or personal income (see e.g. 25, 28–9, 59).

[20] It is an interesting historical irony that payment in kind and wage supplements assisted in maintaining domestic demand for commodities like cotton textiles, manufactured with new industrial techniques, using a new style of industrial labour. The success of these industries in turn transformed the social and economic structure of the society that nurtured them, gradually eradicating the old practices.

awkward for economic historians; but they were intrinsic to the functioning of society, giving a flexibility too-little acknowledged by modern observers.

Another simple strategy to counter declining real wages was to work for longer hours in preference to greater leisure and diminished income. The newly mechanized trades, by their nature, involved longer regularized hours of work of twelve to sixteen hours for six days a week, with no optional 'Saint Monday' holiday if labourers had over-indulged on a Sunday. It has been accepted that greater effort and longer hours were features of much of the employment of this period. O'Brien and Engerman described this as 'an increased labour output in the market sector', equating the increase in hours with a corresponding greater demand for goods in the home market. They further related that,

debates on the standard of living and on the causes of the industrial revolution have been intertwined. A pessimistic position on the former, since it presumes inadequate domestic demand, is consistent with an emphasis placed upon the role of foreign markets in generating growth. Recent work on foreign trade patterns and upon the sources of demand, demonstrates, however, that the home market was a major factor in the demand for increased production, and that the home market must have accounted for much of the increase in output demand in this period.[21]

The reshaping of work patterns contributed substantially to the health of family incomes and so to the demand for consumer goods. Fundamental alterations in the structure and composition of wage-labour occurred during this era, chief of which was the paid employment of women and children. Neither women nor children had existed in idleness prior to this. Their payment for work, however, was either very small or irregular. The common element of most of their work was the lack of cash payments, without which this enormous section of the population would be forever outside the mainstream of consumers. This dependent economic position barred them from readily making personal choices as consumers, impeded them from any positive contribution to demand aside from the necessities of life. As long as this situation persisted the home market was substantially diminished.

Women and children joined in the wage economy in earnest with the proliferation of the domestic system, which saw the manufacture of an

[21] O'Brien and Engerman, 178.

expanding variety of goods from the late sixteenth century onwards. Long before the technological innovations that led to the factory system, cottage industry enabled women and children to contribute cash earnings to the family income. Early in the eighteenth century, Daniel Defoe drew a vivid contrast between the lot of the agricultural labourer's family, dependent on one wage of twelve shillings, and the family dwelling in a textile-manufacturing district. In the latter instance, the father's income was matched by that of the family. The wife would spin as too would the older daughters, while the smaller children worked for the weaver, winding weft and performing other tasks for small sums. The benefits of this system of work, as Defoe described it, were that 'they all feed better, are cloth'd warmer, and do not so easily fall into Misery and Distress'. Defoe emphasized the particular effect that the salaried woman had on the consumption of goods within the family, stating that whereas 'the Father gets the Food ... Mother gets them Clothes'.[22]

Children were customarily put to work at a very early age under the domestic system of production. One of Samuel Crompton's earliest memories was of being stood up in a washing tub of soapy water with layers of cotton battens under his feet, which he was to wash as he trod on them.[23] William Radcliffe described the jobs he performed as a child when his mother 'taught me to earn my bread by carding and spinning cotton, winding linen or cotton weft for my father and elder brothers at the loom, until I became of sufficient age and strength for my father to put me into a loom'.[24] The importance of the expansion of consumer industries, both new and old, lay in the additional and independent sums paid to women and children. This point has been dealt with in detail by Neil McKendrick.[25] The prosperity of the cotton industry, and of the many other trades which flourished in the eighteenth century, depended on the health of national demand; this, in turn, enabled a new segment of society to find wage-earning

[22] Daniel Defoe, *A Plan of English Commerce* (1728, repr. 1927), 69.
[23] Ivy Pinchbeck, *Women Workers and the Industrial Revolution, 1750–1850* (1930, repr. 1981), 113.
[24] William Radcliffe, *History of Power Loom Weaving* (1828), 9–10, quoted in Pinchbeck, 113. It is beyond the scope of this work to examine the conditions under which these children worked in the domestic system, but it is apparent that with respect to unpleasant, dangerous, and unhealthy conditions, some situations of domestic labour rivalled the early factory system.
[25] McKendrick, 'Home Demand and Economic Growth: A New View of the Role of Women and Childrn in the Industrial Revolution', in Neil McKendrick (ed.), *Historical Perspectives* (1974).

employment and to participate as buyers in the market-place. The opening of spinning-mills, the expansion of the potteries, and the proliferation of workshops devoted to the production of ornamental smallwares presented the opportunity for employment of a profoundly new kind. Hours were long, but the wages obtained were far in excess of the two-and-a-half shillings which was the average earned by a cottage spinner. Moreover, as Neil McKendrick notes, long working days were not new for women, 'but the machinery to make these hours of work as productive as the work of men was new'.[26] The addition of these earnings to those of the husband made a qualitative difference to the levels of surplus income in the family. McKendrick calculated that if wage-rates for men, women, and children were 3 to 2 to 1, or even 4 to 2 to 1, the total weekly income of a working family would, at least, almost double that of the husband's alone, and at best equal over twice his weekly salary.[27]

The cotton industry thrived in the provision of all manner of knitwear and textiles, leading to the recruitment of a virtual army of spinners and auxiliary trades in the cottages of rural Cheshire, Lancashire, Yorkshire, Nottinghamshire, and elsewhere. The recorded imports of raw cotton (see Table 2.4) provide an indication of the growing numbers working in this trade, as well as the expanding numbers of wage-earners now more able to participate as consumers in the home market. The volume of cotton may appear insubstantial in comparison to the imports of the later machine-spinning age, but those are not the figures against which it should be compared. Imports of raw cotton were calculated at 2,325,000 pounds in 1746 and reached 6,687,000 pounds in 1766.[28] Viewed as a reflection of manual spinning hours, the increase in retained raw cotton of over 40 per cent from 1750 to 1774, is a striking testimony to the growing numbers of adults and children who found employment in the domestic production of textiles. The contribution of women to this trade was apparent, as was the projected earnings of these spinners. Along with the advance in cotton imports was a parallel growth in the ability of these labouring families to participate more easily in the developing consumer society.

Heretofore women and children figured hardly at all in the calculations of consumer demand in the national market. This contribution to demand would have been marginal without an

---

[26] Ibid. 164.   [27] Ibid. 184.
[28] Elizabeth Schumpeter, *English Overseas Trade Statistics, 1697–1808* (1960), Table XVIII, p. 62.

TABLE 2.4 *Retained imports of raw cotton* (1,000/lbs)[29]

1720–4	8,519	1760–4	12,807
1725–9	7,576	1765–9	22,008
1730–4	7,443	1770–4	18,392
1735–9	10,012	1775–9	29,380
1740–4	8,320	1780–4	43,143
1745–9	11.142	1785–9	104,225
1750–4	14,739	1790–4	126,908
1755–9	12,786	1795–9	152,244

independent income to effect purchases. But as the eighteenth century progressed, this group had money to spend and the sum total would figure significantly in the national equation. For the first time hundreds of thousands of women and children received regular wages, while in the associated trades working women also achieved a small prosperity. Women working as carriers, shopkeepers, warehouse operators, and in a variety of other retail occupations were to be found in this period, many of whom appeared to be enjoying a prosperity founded on the expansion of the cotton industry and other consumer trades.[30] During the first decades of the century the long series of good harvests and expanding agricultural production fed consumer demand largely through the advance of the purchasing power of men's wages. Neil McKendrick points out, however, that

> the entrepreneurial success stories of the late eighteenth century demanded more dramatic increases than these, and before the appearance of large-scale exports, the increased family unit earnings, swollen by the wages of wives and children, played a major part in providing [the surplus to spend].[31]

---

[29] Figures based on the statistical table devised by Elizabeth Schumpeter, ibid. This table excludes the year 1749, for which figures are not available.

[30] Women listed as carriers appear in the *Universal British Directory* (1793), iii, 668, operating short hauls on the Manchester, Warrington, Ormskirk routes. Women shopkeepers appeared regularly in the Day Book of Sales of the unknown Manchester firm, 1773–9, MC MS: ff 657, D43, Manchester Public Library. Women warehouse 'men' appeared in H. Lowndes, *A London Directory* (1776), 98, and in *Pope's Bath Chronicle*, Sept. 1766. These are simply a sampling of the sorts of employment and businesses undertaken by women at this time. For a later survey of women's occupations well after the onset of the Industrial Revolution see Pinchbeck, Appendix, based on the Occupational Abstract of the Census Returns, 317–21.

[31] McKendrick, 'Home Demand and Economic Growth', 195–6.

A large family was no longer necessarily a burden, or the recipe for hunger and want. Now these children could be fruitfully employed and, with their mothers, add to the material prosperity of their families.

During the eighteenth century, women assumed a more prominent role as consumers, making the choices as to what would be bought. Defoe in his time, and McKendrick, more recently, both pointed out how demand was heightened and the market expanded by the addition of tens of thousands of women as wage-earners. Women's function within a household predisposed them to be consumers of clothing and furnishing textiles. Women made decisions within the family about the replacement of old or outmoded clothing, the addition of soft furnishings, and the provision of household linens. Once they are in the position of wage-earners they advanced as more active consumers. Records from the Greg Mill at Styal show how readily the female operatives transformed their improved economic standing into a rise in demand for various consumer goods, among which were new cotton clothing. Fashionable clothing was the first purchase in times of plenty and, along with hats and calicoes, these women selected various practical garments and household linens.[32] Women's role in making consumer choices, as a new agent in national demand, is a significant component in its profile.

The importance of expanding middle-class demand in the progress of the industrial consumer trades hardly needs reiteration. It is worth mentioning, however, that the ranks of the improving working classes were also contributing to national demand. Among hundreds of thousands of these families, British cottons found their largest market. Periodic fluctuations in the economy and sporadic depressions and dislocations in trade occurred intermittently during the century. Notwithstanding these checks, the most constant factor was the continuing positive expectations of the economy, evinced by the population as a whole and reflected in the declining age of first marriage and sustained rise in the birth-rate. Paid employment in domestic industry enabled women and children to add to the family budget an amount equivalent to the father's earnings; payments in kind, perquisites, and related strategies buffered the family, strengthening demand at the broadest levels in the nation as a whole.

---

[32] The Quarry Bank Papers, Styal Mill Shop Accounts, quoted in Francis Collier, *The Family Economy of the Working-Class* (1964), 40–1.

CULTIVATING POPULAR DEMAND

The eighteenth century was characterized by a constant and growing preoccupation with material betterment. Lorna Weatherill's extensive research into the ownership of goods, as reflected in probate inventories, suggests substantial changes in consumer patterns among those with sufficient property to leave such records. A sampling of several thousand probate inventories from eight areas of Britain indicated, for instance, that the proportion of people owning window curtains rose from 7 per cent in 1675 to 21 per cent in 1725.[33] Later results from this study disclosed that by 1725, 11 per cent of male and female probated wills owed china and 18 per cent had appropriate pottery for hot drinks.[34] All of these innovations in consumer demand were devoid of any element of necessity. Yet the power of these desires among the population became an engine for economic growth.

The potency of popular demand for clothing, china, or domestic accessories cannot be overestimated, and entrepreneurs were quick to appreciate the commercial potential of broadly based national fads. Manufacturers strove not only to produce the merchandise but also to feed the demand for these goods through the purveyance of information and the dissemination of news in print. In turn printers multiplied, flourishing with the provision of a new sort of advertising. Newspapers, journals, and magazines sprang up to feed the town, the region, or even the nation with the latest news and, equally important, relayed the details of the latest products to the public. Shopkeepers described their merchandise in glowing terms, announcing their stock of fabrics to be 'sold as cheap as in London' or 'of the widest assortment outside London'. Custom was solicited and fresh items heralded, as, for example, the availability of new copperplate printed cottons in Bath.[35] Advertising, whether by handbill, trade-card, or newspaper, brought the expanding range of merchandise to the attention of the public in a more efficient manner that was without precedent. Previously customers in the vicinity of a stall or shop-front

---

[33] Lorna Weatherill, 'Consumer Behaviour, Textiles and Dress: The Evidence from Probate Inventories and Household Accounts, 1670–1730', paper, Pasold Conference on the Economic and Social History of Dress, 1985, pending; id. 'A Possession of One's Own: Women and Consumer Behaviour in England, 1660–1749', *Journal of British Studies*, 25 (1986), 139.

[34] See also, Lorna Weatherill, *Consumer Behaviour and Material Culture in Britain, 1660–1760* (1988). [35] *Pope's Bath Chronicle*, 7 Mar. 1765.

might be enticed to enter and examine the stock through the shouts, cries, or appeals of the traders. But such solicitations were restricted to passers-by, neighbouring inhabitants, or those attracted to a clustering of shops or stalls. Word of mouth might attract patrons and tradition brought customers to large established markets, but essentially most dealers served only the immediate locale. The methods of drumming up custom were haphazard at best, and little information could be included in the shout or call to lure the shopper to buy. Printed advertisements brought to the attention of the reader all the essential facts about any number of trades: goods could be listed with prices and new stock announced, making this information available systematically to thousands of potential customers. Handbills, trade-cards, and newspapers brought a visual image and factual information to those able and willing to buy. (See Plate 6, the trade-card for Cotes's Manchester Warehouse.)

The efficiency of this sort of advertising can be gauged from local and national papers. In a local paper like *Jackson's Oxford Journal* advertisements were placed regularly by the principal drapers, mercers, and hosiers of the town, while periodic notices were also placed by tradesmen from the surrounding counties of Buckinghamshire, Worcestershire, and Berkshire.[36] The extended distribution of regional newspapers assisted directly in forging the commercial links of the nation relaying accurate commercial information—both retail and wholesale. Provincial gazettes developed their distribution systems early in the century; beginning with weekly sales at the corn-markets and progressing to routine deliveries to the homes of subscribers by 1720. Direct dispatches to surrounding towns, villages, and countryside were soon considered indispensable by local corporations who sanctioned the establishment of local papers. Stamford Corporation made this a prerequisite for the two printers who wanted to set up business in that town. Over the next decade local news-sheets, such as the *Reading Mercury*, boasted a distribution of forty or fifty miles around Reading. Even this level of improvement did not satisfy the printers. Shortly thereafter ambitious printers were arranging for local agents to organize even more extensive sales networks, and by 1743 the *Reading Mercury* listed thirty-six agents as distributors, operating within a circumference that included Rye, Portsmouth, Salisbury,

---

[36] *Jackson's Oxford Journal*, 18 Apr. 1767; 2 May 1767; 2 June 1770; 20 Jan. 1770; 14 Apr. 1780; 16 June 1781; 23 June 1781.

Bicester, Aylesbury, and London. The distribution of the *York Courant* was similarly impressive, as were numerous other provincial papers.[37]

Advertisements were sought from local merchants, plus any London-based firms who recognized the value of advertising in provincial papers. Thus, in 1743, the *London Evening Post* carried a notice from the *Reading Mercury* announcing that,

> This is to give *Notice*, to all Persons who may have Occasion to advertise in the South and West Parts of England, that the *Reading Mercury* ... is the most convenient Paper for that Purpose, it being distributed throughout Berkshire, Buckinghamshire, Oxfordshire, Wiltshire, Hampshire, Surrey, Sussex, and part of Kent and Middlesex. ... N.B. Gentlemen who live in London may see that their Advertisements are properly inserted, if they apply to Mr John Kemp at Sam's Coffee-house in Ludgate-street ...[38]

The *Liverpool Chronicle* devised a more novel strategem, forwarding free copies to 'the principal Coffee-Houses and Taverns', to bring the paper and its contents to the attention of potential customers and advertisers. The final development came with the decision of the *Bristol Oracle* to use the post office as the basis of its distribution system, promising to its advertisers that, 'Advertisements for this Paper, which has the singular Privilege of being freely circulated throughout all the Cross-Roads in England, may be sent by Post, without any Expense'.[39] These provincial newspapers drew advertisements from throughout their zone of distribution and in turn spread information, conveying, even to the remotest areas of England, notices of the rising momentum of commercial life. One reader of the *Derby Mercury*, living in 'a Corner of the High Peak of Derbyshire' wrote the editor that, 'though recluse and at a great Distance from you ... yet your Paper reaches me Weekly by the Packhorse Track'.[40] Eventually newspapers permeated all regions of the countryside, bringing with them news of goods for sale and bargains to be had to tempt the shopper and inspire the local shopkeeper to greater heights of mercantile activity.[41]

London tradesmen were equally assiduous in their attention to advertising, with the intent of making their names and businesses

---

[37] G. A. Granfield, *The Development of the Provincial Newspaper, 1700–1760* (1960), 190–8.
[38] *London Evening Post*, 31 Mar.–2 Apr. 1743, quoted in Granfield, 199–200.
[39] *Bristol Oracle*, 25 June 1743, quoted in Granfield, 200–1.
[40] *Derby Mercury*, 25 Jan. 1754, quoted in Granfield, 204.
[41] Granfield, 190–2.

known to all readers of the newspapers. A perusal of contemporary papers shows that many of the largest retailers advertised their merchandise regularly and apparently to great benefit. Thus Mr Taylor of London owned two warehouses selling quilted petticoats, when in January 1779 he announced the opening of a third. This style of advertising was by no means new; London tradesmen had long used the printed medium to parade their wares. But by the second half of the century momentum in advertising grew. Linen warehouses, Manchester warehouses, shirt, petticoat, cloak, and haberdashery warehouses advertised regularly, as did tailors, milliners, mantua-makers, and the like. All aimed at convincing the reader that here were things whose novelty, utility, and price made them irresistible.[42] Such was the success of this new venture that journals sprang up bearing the word 'advertiser' in their title. *The General Advertiser* operated beginning in the first half of the century, and it, as well as the *Public Advertiser* and *Gazetteer and New Daily Advertiser*, clearly intended to attract readers with the promise of plentiful and accurate commercial intelligence. In 1760 Londoners had morning and evening newspapers to choose from, as well as several weekly publications. Twopence a week could buy a weekly paper.[43] The channels of distribution were so extensive by this time and the papers so numerous, that no one was obliged to be ignorant of markets or commodities. Moreover, by the end of the eighteenth century the simple advertisement had been joined by more sophisticated appeals to consumers, more eye-catching stimuli to demand. In the autumn of 1782 one large London clothing emporium commenced its advertisements with the claim of 'Wonder of Wonders! and the Greatest of Wonders'; a marked departure from the stark listing of textiles only decades before.[44] Richard Sheridan's Mr Puff in the 1779 play *The Critic* typified the bumptious, pushing advertiser, one of the new category of thrusting men who worked at systematizing the outrageous claims for goods, seeding acquistiveness, and helping to propagate demand.

The unique advertising techniques of one George Packwood, a trader in shaving accessories, were uncovered by Neil McKendrick. To what heights would this tradesman not go to entertain, intrigue, and attract his customer? The advertising campaign launched and

---

[42] *The Gazetteer and New Daily Advertiser*, 25 Feb. 1777; 27 May 1777; 22 Aug. 1777; 27 Oct. 1778; 2, 14 Jan. 1779.
[43] J. H. Plumb, *Georgian Delights* (1980), 13.
[44] *The Gazetteer and New Daily Advertiser*, 4 Nov. 1782.

sustained by this lowly entrepreneur, from 1794 to 1796, included puzzles, stories, testimonials, dialogues, and riddles striving to make the name of Packwood synonymous with shaving. Dr Johnson's claim in 1759 that 'the trade of advertising is now so near perfection that it is not easy to propose any improvement' was somewhat premature.[45] Certainly advertisements had become a commonplace by that time, and so more extravagant claims and attention-grabbing headlines had to be incorporated in the text to obtain the desired results—that the eye of the reader be focused, however briefly, on the content of the advertisement. By the last decades of the century advertisements such as that noted above would not have been in the least extraordinary, particularly when compared to the creative acumen of George Packwood, who used all manner of slogans, jokes, riddles, and letters singing the praises of Packwood's razor strop and balm.[46]

These innovative advertisements and their duller ubiquitous counterparts were signals: flags raised in praise of consumerism, demanding the attention of the readers of newspapers, urging further expenditures; setting the standards of material sufficiency ever higher. (See Plate 7, inventive advertising by Romanis, Hosier.) Such advertisements are also signals to historians, as Neil McKendrick points out, announcing qualitative changes in the patterns and rate of sales, innovations of comparable significance to the changing patterns of production.

In overlooking the collective importance of these forgotten men of business historians all too often seriously distort the historical record, and greatly over-simplify the process of economic change. ... Packwood and his retailers selling his goods under the sign of the Naked Truth were essential agents in the process of commercialization. Such men provided the network of commercial outlets which helped to make possible the emergence of a consumer society. Without their assiduous advertising, without their persistent promotion of their product image, without their remorseless repetition of their brand names, aggregate home demand would have lost a vital sustaining element in its growth and its social and geographical diffusion. They spread both news of their products and the products themselves. They made them both fashionable and accessible. They made them more desirable and more easily purchasable. They were, in H. G. Well's phrase, the true 'propagandists of consumption'.[47]

---

[45] Dr Johnson, quoted in Neil McKendrick et al., *The Birth of a Consumer Society* (1982), 151.
[46] Ibid. 152–3.
[47] Ibid. 192, 194.

From one end of the country to the other newspapers and journals advised their subscribers of the new items available, among which would be the expanding assortment of cotton textiles. Prices, varieties, and new innovations in stock were all related. Information was a spur to fancy: fashion and fancy stimulated demand. Throughout the century articles were written and sermons preached about the deplorable tendency towards excessive materialism witnessed in all levels of society. Yet the expectation of an improving material existence persisted and remained an ever-present, though immeasurable, constituent of economic life and domestic demand. A moralist wrote in *The Lady's Magazine* that he was amazed to see that 'young people spend so much of their precious time in decking these poor frail bodies, which . . . will last but a short season and then must be prey for worms and moulder into dust'.[48] None the less, such temporal indulgence was the staff of economic life; asceticism was never a priority of the growing generations of young people living in the later eighteenth century. Far from decreasing, public fascination with the new commodities persisted, as the producers, retailers, and printers combined their efforts to foster demand. Judicious advertising had the same effect on the domestic market as the liberal use of manure had on farm-land over the same period. Returns were improved, standards of comfort raised, and the country enriched.

HIDDEN DEMAND: THE SECOND-HAND TRADE

The demand for new clothing, textiles, pottery, metal-ware, and other consumer goods extant in Britain was not the total sum of the consumer impulse. An equally powerful market-demand was manifested not through the purchase of new commodities, but through the sale, trade, and purchase of second-hand merchandise. Joan Thirsk has noted that 'the labouring classes found cash to spare for consumer goods in 1700 that had no place in their budgets in 1550'.[49] But among these mass of people in the middle and lower strata not all demand for novel or attractive articles was translated into the purchase of new goods. British men and women routinely assuaged their needs and wants with the purchase of used merchandise. Demand was two-tiered. At the top was the open and apparent consumer demand

---

[48] *The Lady's Magazine*, 5 (Jan. 1774), 14.
[49] Joan Thirsk, *Economic Policy and Projects* (1979), 174–5.

expressed by the middle and upper ranks through acquisition of new furnishings, clothing, tableware, and the hundreds of miscellaneous items manufactured in growing numbers in the workshops of Britain. Beneath this lay the most numerous of Britain's families, with an income of less than the £50 per annum suggested by D. E. C. Eversley as the minimum that would enable intermittent participation as a consumer. These families were not barred from participating in the round of consumerism that occupied the wealthier classes, although they did not have as much to spend and were limited in the range of new articles they could afford. Throughout Britain there also existed a well-established, organized system of redistribution, founded on the demand of those in more straitened circumstances. The trade existed because the needs of the whole population could not yet be met within the existing structure of production. The identification of the characteristics of this second tier of demand are crucial for an understanding of the British home market in the eighteenth century. The second-hand trade was a key intermediate trade, using barter as well as cash sales, in the movement of goods through the nation. The influence of this largely hidden trade and of those who sustained it are fundamental factors at work in the development of the cotton industry and in the diversification of its products to meet the needs of the whole of Britain's population.

Undoubtedly the second-hand trade existed, at least in major centres, for generations or even centuries before it came to the notice of commentators. Written record of this trade arising in the eighteenth century can be explained by the vastly greater volume of merchandise which it came to handle, in addition to the more visible numbers of traders and their patrons. The second-hand trade developed as a source of substitutes, enabling millions of lesser folk to make do with second-hand as long as the cost of new materials kept those items out of their reach. The scope of the second-hand trade was dependent on the time it took for industrialized production to lower costs sufficiently to offer fashionable new clothing to the mass of the population at prices they could afford. This imperative helped determine the manner in which the cotton industry developed.

Records of this trade are limited. Mention of the trade or traders often happened by chance, for, as no manufacturing processes were involved it generated no known response from governments and had no vocal adherents pleading its case to parliamentary committees. The second-hand trade was a commonplace to people of the eighteenth century, requiring no explanation, accepted as a familiar component of everyday life. The frequency with which people of the day would have bought used garments varied with their circumstances and the availability of the required commodities, two aspects which cannot be measured with exactitude. Used apparel was frequently the most practical alternative, providing the poor with cheap covering and offering the ambitious or the more prosperous with the opportunity to dress in clothes that bespoke a higher station.[50]

The sector of society most likely to have bought second-hand regularly was the amalgam of ranks that by Gregory King's calculation accounted for the largest segment of the population—those earning under £50 annually. The appearance of clothing would be important to these men and women, as too would be the cost of a garment; both requirements could be met through the purchase of a second-hand article from a clothes-broker, pawnbroker, itinerant hawker, or local salesman.[51] Second-hand clothing was sold by specialist dealers, as

[50] In the diary of Francis Place he reveals the intriguing preoccupation of both he and his wife Elizabeth with clothing—keeping, maintaining, and acquiring a good stock of clothes, in the early years of their marriage in the late eighteenth century. During the ups and downs of their early married life prosperity was measured by the numbers of clothes they possessed. Unemployment meant the sale or pawning of clothes and prosperity, with renewed employment, found both of them working as hard as they could to build up their supply of clothing yet again. Prior to their marriage Place was calculating their combined assets and contemplating the future, writing: 'As my intended wife had a good stock of cloaths and I hoped that I should be able by working hard to increase my very scanty quantity, I thought we could go on well in this respect.' During a period of economic stasis Place recalled that, 'We soon acquired the character of being proud and above our equals, this was the certain consequence of our having no acquaintance with any one and being better dressed than most who were similarly circumstanced, and were contemptuously called *the* Lady and Gentleman'. Later a period of intensive work brought its rewards in the material improvement to the couple's wardrobe and surroundings. 'I worked incessantly, and soon saved money enough to buy some good cloaths and a bedstead, a table three or four chairs and some bedding, with these and a few utensils we took an unfurnished room up two pairs of stairs at a chandlers shop ... and began to congratulate ourselves on the improvement of our circumstances and the prospect before us.' While there is no mention where and by what means clothes were bought, their mutual attention to dress perhaps bespeaks the attitude common to the ambitious journeyman and suggests as well the source of part of the demand that channelled into the second-hand trade. Mary Thale (ed.), *The Autobiography of Francis Place* (1972), 102, 106, 111.

[51] 'Salesman' was a term routinely used to describe someone who traded in clothing, usually used, though new goods could also be found in their inventory. R. Campbell, author of *The London Tradesman* ... (1747), wrote in his summary of London trades that: 'Salesmen deal in Old Cloaths, and sometimes in New. They trade very largely, and some are worth some Thousands'. Other terms used to describe tradesmen of similar interests were sometimes peculiar to geographic regions: for example, in

well as by many other traders large and small, enabling untold thousands to wear clothes above their means had the garments been new. Moreover, the slight discrepancy in fashion between London and Lancashire, Exeter or York, would not inhibit the second-hand trade in the least. Broadly speaking the English style of dress was uniform throughout the country. Gradations in apparel were more directly tied to the ability to afford costly fabrics, finishing, and accessories; differentiations in dress would then accord with rank, access to the metropolis or other sources of information, and appropriateness to local circumstances. The latest London dress was not always appropriate in rural communities, but clothes a year or two old would not offend. Thus, clothes outmoded by the calculations of one group would be in demand and thought desirable by another. The time-lag in fashion diminished gradually over the course of the eighteenth century. Fashions became more generally uniform across the nation and the classes of society. But long before the desire for fashionable clothes could be completely met through the production of new materials, this demand stimulated the purchase of second-hand clothing throughout England.

Much of the trade in used clothing remains uncharted, though some points of this process are well known and well documented, such as the making of routine gifts of clothing and linens to servants and the lively clothing trade along Monmouth Street, Rosemary Lane, and Petticoat Lane in London.[52] For the most part the trade left few records, attracted little attention, and left only a poorly marked trail for historians. Occasional mentions of this trade surface in contemporary newspapers or trial reports; suggestive notations appear sporadically among shopkeepers' records. But in general the trade was of interest only to those with something to sell or to buy.[53] It was a vital conduit for the people of this era, both as an avenue through which they could barter or sell their used items and as a source of inexpensive garments

---

Cheshire a dealer or broker in household goods implied also a trade in clothing and furnishings. A nineteenth-century water-colour by Louise Rayner of the Water Street Row, in Chester, depicts the premises of just this sort of trader, with the requisite pieces of clothing and furnishings before his door.

[52] Madeleine Ginsburg has written what is probably the first modern appraisal of the London second-hand clothes trade in 'Rags to Riches: The Second-Hand Clothes Trade, 1700–1978', *Costume*, 14 (1980), 121–35.

[53] Note in 1763, James Boswell reporting the sale of an old suit and lace from his hat, the proceeds from which were used to stretch his finances to cover the expensive pursuit of Louise. Fredrick A. Pottle (ed.), *Boswell's London Journal, 1762–1763* (1950), 115.

of every sort, at every price. The mountains of gowns, jackets, aprons, stockings, and breeches brought to the salesmen or brokers, were not seen as valueless, fit only for charity, but as articles of varying worth that would bring a profit to the trader and add to the assortment of clothing available to the consumer. The trade in used clothing and textile items developed precisely because of the value and utility of these items, the uniformity and homogeneity of the home market, and the high level of demand at all levels of society. This trade operated on the fringe of the textile and clothing industries and began where the involvement of all first-phase manufacture and sale ended, after the consumer had bought and worn her new clothing and then, for whatever reason, decided to sell it. When fashions changed, when fortunes waned, or when new clothes became imperative, tradesmen were ready to buy the soiled, shabby, or *passé*, redistributing them in a trading network that spanned Britain, her colonies, and Europe.

Britain was well served by retail tradesmen, whether they were chapman or shopkeepers, while many fairs and markets continued to function as additional centres of commerce throughout this period. T. S. Willan's *Inland Trade*, and more recently Margaret Spufford's *The Great Reclothing of Rural England*, have uncovered the extensive interwoven grid of middlemen that operated in the early modern period, wherein chapmen, retail shops, and fairs all served to distribute goods among tradesmen as well as carry goods directly to consumers. Domestically manufactured products reached every corner of the British market by the beginning of the eighteenth century. Nevertheless, in past considerations of the market and trade in second-hand clothing it was assumed that this trade would be localized in London and other provincial centres.[54] As retail distribution spread throughout Britain, so to would tradesmen profit from the unwanted articles of an increasingly prosperous society, redirecting apparel to satisfy the demand of a less affluent segment of the population.

Tradesmen's records suggest that the role of the local shopkeeper in the second-hand trade was slight—an intermittent function of their business. In June of 1762, for example, the Cumbria shopkeeper Abraham Dent listed the sale of 'an old Coat' to James Petty, a labourer on the local turnpike.[55] But aside from the sporadic sale of used apparel, shopkeepers in the provinces may also have participated

[54] Ginsburg, 'Rags to Riches', 122.
[55] WDB/63/3a, Cumbria Record Office, Kendal.

in the collection of used textiles from their customers. The recently published *Diary of Thomas Turner* examined the very wide range of business activities engaged in by Turner as a shopkeeper in a small village in Surrey from 1754 to 1765. Turner collected what he called 'rags'. He does not elaborate on what constituted rags, but it can be assumed that these would include all sorts of redundant textiles collected from households that would be of value to others. In 1759 Turner made two collections of rags. He packed '2 cwt. 1 qr. 26 lb. of rags' to be taken by his agent to the Maidstone Fair, while earlier that year he had sold other bales of rags directly to the paper-makers.[56] It was common for rag-gatherers in the country to deal directly with the paper-makers.[57] Moreover, it is apparent both from Turner's diary and the history of paper-making that the collection of rags—used textiles—was a normal aspect of the circulation of textiles. But would the five and six hundredweight of rags collected by Turner over the next year all go to the making of paper? It seems far more probable that there was a differentiation made of these 'rags' enabling traders to profit from the single items of used clothing which would overall bring in much more than the thirty-five shillings per hundredweight, the standard price for rags in the 1780s.[58] A good second-hand coat could cost five shillings and a poor one as little as fourpence,[59] but the sorting out of this sort of 'rag' would bring the independent tradesman more profit than selling serviceable clothing by weight to be broken down into pulp for paper. One must assume that eighteenth-century tradesmen were at least as diligent in the search for profits as their modern counterparts and that they too would put into effect strategies still employed in the processing of rags and used clothing.

Certainly paper-makers serving the London market used quantities of rags in the production of brown and white paper; but paper-making was by no means the only outlet for cast-off garments, particularly when there was still life and wear left in the clothing. The London and national market as a whole exerted a strong demand for used clothing. Some of the 'rags' sold by Turner at the Maidstone Fair might well have been sorted and resold to dealers in wearing apparel as well as to paper-makers. These suggestive sales by Turner require further investigation into the recycling of textiles by town and village

[56] David Vaisey (ed.), *The Diary of Thomas Turner, 1754–1765* (1985), 185, 190.
[57] D. C. Coleman, *The British Paper Industry, 1495–1860* (1975), 37.
[58] Ibid. 173.
[59] Ginsburg, 'Rags to Riches', 122–5.

shopkeepers before categorical conclusions can be arrived at about the function of intermediate retailers in this trade.

Evidence does exist of tradesmen operating nationally, buying and reselling wardrobes and used clothing. The gentry and middling ranks bought clothes for reasons other than necessity, and when no longer needed these clothes found their way back on to the market. John Matthews advertised his willingness to buy just these sorts of goods in *Jackson's Oxford Journal* in the spring of 1770.

John Matthews, Salesman from London, buys Ladies and Gentlemans castoff Cloaths, either laced, embroidered, or brocaded, full-trimmed, or not, of every Colour and Sort, and will give the most Money for any: As I can deal for London, the Country, and Abroad, nothing can be out of my Way, according to the Price and if any Person has any thing to dispose of and will favour me with the Sight of it, they may depend on having the full Value of their Goods.

Matthews planned to cover parts of Berkshire and Buckinghamshire, if required, as well as Oxfordshire, stating that he would travel up to twenty miles out from Oxford during his two-week stop in the region. This London salesman traded in all manner of clothing, accessories, and linens. He announced that, 'I likewise buy all Sorts of old Linnens, Gold and Silver Lace, burnt or unburnt, School Boys Cloaths and Servants Liveries'.[60]

Aside from the wealthy specialist tradesmen many small dealers participated in the collection of second-hand clothing. The process of accumulating used apparel described in Henry Mayhew's *London Labour and the London Poor* in the mid-nineteenth century, was one that had been repeated among pedlars and traders for generations. Mayhew recorded the practice of the crockery sellers whose trade involved the exchange of new goods for old, a process as old as the tale of Aladdin's lamp. The crockery sellers walked their routes around London and its environs, crying 'any old clothes to sell or exchange'— a cry that had been familiar to residents of London for centuries. The crockery sellers carried baskets weighing about seventy-five pounds and 'travelled about fifteen miles with that, at least; for as fast as I gets rid on the weight of the crockery, I takes up the weight of the old clothes'.[61] By the end of his route, if all of the crockery had been sold then he no longer exchanged but bought as many old clothes as he could. A case before the Old Bailey magistrate in 1742 depicted an

[60] *Jackson's Oxford Journal*, 17 Mar. 1770.
[61] Henry Mayhew, *London Labour and the London Poor* (1851, repr. 1967), i, 368.

identical practice among the earthenware dealers of the period, in which used clothing was accepted in payment for crockery.[62] (See Plate 8, an identical trade noted in a 1740 advertisement placed by Hannah Tatum.)

Even more suggestive was the account of the trade of James Varlow and his wife. Varlow had travelled through Lincoln and a report of his progress was recounted in the *Stamford Mercury* in 1728. His trade was described as 'gathering Rags and his Wife sells Manchester Ware'.[63] Within the trade of that couple are the components of new and old: the one buying old clothes, 'rags', and the other selling Manchester ware, the products of Britain's cotton-linen industry. One can be sure that their customers were offered the opportunity to buy new Manchester tapes, checks, and other sorts of textiles either through cash payments or the exchange of old goods to the husband for new products from the wife. The exchange of new for old persisted through the Industrial Revolution as a remnant of an older barter system; an antique appendix to a rapidly changing economic structure, but still of use in this intermediary period. The persistence of this method, like the non-cash payments and perquisites among employees, extended the capacity of the common people to participate in this advancing consumer society. Thus a demand for articles among the lower-income ranks could be met through non-cash exchange or payments[64]—an ancient system or barter maintained and adapted to the needs of the pre-industrial and early industrial society. The hawkers, peripatetic dealers in rags, and the like, bridged the cash-based system that was becoming the norm and brought a greater range of products within the reach of common people, putting a significant level of purchasing-power within their grasp.

The collection of clothing was undertaken across the nation, but

---

[62] *Proceedings on the King's Commission of the Peace . . . for the City of London and . . . for the County of Middlesex; held at . . . the old Bailey* Dec. 1742, 14.

[63] *Stamford Mercury*, 25 Apr. 1728.

[64] Non-cash payments not only freed wages for purchases in the market-place by supplementing income, but they could also be translated into cash. The payment of the foreman of the tailors in Wild's and Place's establishment is a case in point. His wages were two-and-a-half guineas per week, plus his clothes and other unspecified perquisites. Payment in clothes and even the raffling of a new pair of breeches at a Breeches Club, such as Place established, were popular in a way that is surprising by modern standards. But it must be remembered that the winner of the breeches or the payment in kind to the foreman did not only have value in its use; all these items could be turned into cash and were themselves almost a currency, in the way certain commodities still are today. *The Autobiography of Francis Place*, 106–7, 211.

London was the hub of the second-hand trade. The metropolitan requirements of fashion were an asset to the second-hand clothes trade. Tailors regularly sold their client's superfluous garments and with the proceeds produced new goods at reduced cost for the customer. (See Plate 9, placed by Price & Co.) In 1765, William Littlemore, a tailor, proposed in his advertisement a sort of trade-in on the purchase of a new suit of clothing. Littlemore stated that, 'any Gentleman that chuses to favour me with their Commands, may save a considerable Sum in the Yer, and on the other hand have three times the Choice'. Realizing that some men might not know 'how to dispose of those Cloaths they never intend to wear more', Littlemore promised that 'this will be a Means of preventing any Loss to them'.[65] If Littlemore was specializing in a high class of customer it is doubtful that he kept the used clothes in his shop. A clothes-broker would have taken these goods off his hands and given him a profit as well, at least enough to cover the allowance made to his clients. The business of 'Brown, Tailor and Habitmaker' advertised regularly in *The Gazetteer and New Daily Advertiser*, and Brown also announced that he accepted old clothes on account for new.[66] (See Plate 10, an advertisement by Brown.) Tailors provided a direct link between the new and the old. Occasionally this line was very fine, as Francis Place complained. 'How often have I taken away a garment, for a fault which did not exist... How often have I been obliged to take back a garment and sell it to a jew for not much more or any more than one third of its price, because a man or his wife or his mistress disliked it when it was made up'.[67] The nature of a tailor's business would require some channel to dispose of used or unsatisfactory garments, bringing some sort of earnings back to the business. Thus, discounts on new items were probably available through tailors in towns and cities throughout Britain, simply because a return on the used garments was so assured, whether the tailor resold the goods locally or to the passing wholesalers who toured the country. The refund granted customers on their old clothes encouraged the purchase of new clothing and contributed to the stock of used apparel that would then circulate through the lower levels of society.

The methods employed to collect second-hand clothes are not yet exhausted. Middlemen like John Matthews travelled through Britain

[65] *Public Advertiser*, 28 Jan. 1765.
[66] *The Gazetteer and New Daily Advertiser*, 27 May 1777.
[67] *The Autobiography of Francis Place*, 217.

buying goods as they went, pedlars exchanged new items for old clothes, rag-gatherers and local shopkeepers played their part in the accumulation of stocks of second-hand clothes, and tailors accepted old suits of clothing in part-payment. In addition to these measures pawnbrokers operated as buyers of used clothing. The pawnbroker John Flude promised in his trade-card that, 'Goods Sent from any Part of ye Country directed as above shall be duly attended too & the Utmost Value lent thereon'. Flude dealt not only in clothing, but also in all other valuable items such as plate, jewellery, watches, and the like; but clothing was displayed prominently on his trade-card.[68] Another pawnbroker advertised under the initials J. R.: 'Most Money given for rich and plain Cloaths. Whoever may have any to dispose of, by directing a letter to J. R. No. 6 Plumtree Street . . . Bloomsbury, shall be waited on within ten miles of London'. A promise was further given for those with a position to maintain that, 'Secrecy may be depended upon'.[69] There are undoubtedly many more such advertisements waiting to be discovered in the pages of London's newspapers. As the metropolitan daily and weekly papers were widely read and the capital regularly visited, this was an assured way of obtaining more stock without the necessity of ranging across the countryside.

The ways to dispose of used clothing for the consumer were as numerous outside the metropolis as they were inside. First among the dealers were the pawnbrokers, of whom there was an abundance. As the population of Britain grew in concert with towns, cities, and the volume of consumer goods, the need for pawnshops also multiplied. Labourers, artisans, and servants usually owned few items that could more readily be turned into a cash than their clothing. Pawnshops offered small sums at times essential for a family budget; in exchange for a coat, bonnet, or shawl a vital sixpence might be loaned the customer. Directories of the period provide evidence of the spread of these establishments. *Bailey's Midland Directory* included listings of twenty-three pawnbrokers in Birmingham in 1783. *The Universal British Directory* recorded twenty-one pawnbrokers in Bristol a decade later; Manchester registered eight pawnbrokers; while the centres of Macclesfield, Melksham, Mansfield, Wigan, Whitby, Winchester, Windsor, Worcester, Colchester, Cambridge, Aylesbury, Coventry, Barnet, Bideford, Bridgewater, Bury, Derby, Durham, Atherstone,

---

[68] Trade-card, John Johnson Collection, Bodleian Library, Oxford.
[69] *The Gazetteer and New Daily Advertiser*, 14 Feb. 1777.

Guildford, Reading, Doncaster, Morpeth, Marlow, Margate, Ipswich, Shepton Mallet, Scarborough, Swansea, Taunton, Truro, Canterbury, Henley, Coventry, Leominister, Lancaster, Uttoxeter, Neath, Newark, and Kendal listed at least one each within their precincts.[70] This is only a selection of the men and women working as pawnbrokers, indicating only those pawnbrokers of sufficient stature to warrant inclusion in directories of the age.

Shopkeepers, pawnbrokers, chapmen, and tradesmen all contributed to the collection of second-hand clothing, at the same time providing cash or goods in exchange. Some of the merchandise accumulated would have been transported to London, the heart of the British trading network in used clothes, while the rest would have been dispersed through local or regional distributive networks. Just how many worked at this trade nationally cannot be determined. The collection of second-hand clothes for resale took place on several levels. Through circuitous or direct routes vast stocks of second-hand clothes circulated, a great portion of which were brought to London to be sorted, graded, and resold yet again in a further specialization of the rag trade.[71]

The retail portion of the second-hand trade was equally diverse and it was in this segment that the traders obtained their profits. The sale of gowns, breeches, aprons, waistcoats, and caps, repeated thousands of times over, at market-stalls or tailor's shops, salesmen's stores or London's Rag Fair, was a constituent part of the clothing trade in this period—public demand met by second-hand merchandise. (See Plates 11 and 12, for depictions of perambulatory traders in used attire.) Advertisements for auctions appeared with some regularity in eighteenth-century London newspapers, auctions both for unclaimed pawned apparel and for entire wardrobes. One Mr Burnsall announced the 'Sale of Mrs Moses's Effects', included among which was 'all the rich Wearing Apparel and some other effects of the late Hon. Vice

---

[70] *The Universal British Directory*, vols. i–iv. A total of 644 licences were purchased annually by pawnbrokers at the close of the eighteenth century, according to Patrick Colquhoun's calculations—213 by pawnbrokers in the capital and 431 in the provinces. This would account only for the legitimate followers of this trade. Sir Frederic Eden, *The State of the Poor* (1797, repr. 1966), i, 400.

[71] An indeterminate amount of the clothing added to this assortment was stolen. The theft of clothing figures as one of the principal types of petty larceny and reflects, as well, the pervasiveness of demand for consumer goods. See Beverly Lemire, 'The Theft of Clothes and Popular Consumerism in Early Modern England', *Journal of Social History*, 24: 2 (1990), 255–76.

Admiral Norris, deceased'. Mrs Moses's trade was apparently in high-quality wardrobes and on her death all her stock was auctioned off.[72] 'All the genuine Stock in Trade of the late Mrs Sarah Nowler, Pawnbroker deceased' was also placed up for auction, comprising 'a large Parcel of very good Men and Women's Wearing Apparel', among other things.[73] In other advertisements the garments for sale were noted specifically:

Several Dozen of good second-hand ruffled and plain Shirts, and some Mens laced ruffles, a whole Set of curious China ... with various other Articles, to be sold extremely cheap, at the Acorn, the Corner of New-court, in the Butcher Row, Temple-Bar, where all Sorts of Goods are redeemed out of Pawn to sell, or bought if no pawn.[74]

To the diverse environs of Rosemary Lane and East Smithfield, Houndsditch, the Minories and Petticoat Lane, Chick Lane and Long Lane, the Barbican, and of course Monmouth Street came buyers and sellers, anxious to find bargains or to make a sharp deal with the unwary.

John Strype described many of the locales of London where the trade in second-hand goods flourished during the first half of the century. He depicted the trade and some of those whose trade it was, though little of what he said about the latter group was laudatory. Of Birchin Lane, Strype wrote simply that 'It was a Place of considerable Trade for Men's Apparel; the greatest Part of the Shop-keepers being Salesmen'.[75] The settlement of pawnbrokers outside the city walls along Houndsditch in the sixteenth and seventeenth centuries elicited a much less sanguine description in the telling of the development of this segment of the second-hand trade.

These Men, or rather Monsters in the Shape of Men, profess to live by Lending, and will lend nothing but upon Pawns; neither to any, but unto the poor People only ... let me not here be mistaken, that I condemn such as live by honest Buying and Selling ... No truly, I mean only the Judas Broker, that lives by the Bag, and, except God be more merciful to him, will follow him that did bear the Bag.[76]

[72] *Public Advertiser*, 8 Mar. 1765.
[73] Ibid. 2 Jan. 1760.   [74] Ibid. 10 Jan. 1760.
[75] John Stow, *Survey of the Cities of London and Westminster... Corrected, Improved... in the Year 1720 by John Strype...*, 6th edn. (1754), i, 474.
[76] Ibid. 366–7.

The later mid-eighteenth century descriptions of this area displayed much less spleen and the simple description related only that, 'This Street is a Place of good Trade, and of Note for Salesmen and Brokers, whose Dealings are in Apparel, Linen and Upholsterers Goods, and chiefly second-hand Goods'. The Barbican was called 'a Place of Note for the Sale of Apparel, Linnen, and Upholsters Goods, both second-hand and new, but chiefly for old, for which it is of Note'. Long Lane too was said to be inhabited in the first half of the eighteenth century 'chiefly by Shopkeepers who deal in Apparel, Linnen and Upholsterers Goods, both new and old'.[77] From the accounts of these areas of London it is apparent how readily there was access to second-hand goods. This was a common trade. Its very humbleness precluded much note, either by contemporaries or by historians. In the east, salesmen's premises were noted on Mile End, Whitechapel, Shoreditch, Mansell Street and Prescott Street in Goodman's Fields, Rosemary Lane, the Minories, and of course Houndsditch. At the end of the eighteenth century salesmen continued to congregate in great numbers along Houndsditch; twenty-five such shops were listed in that vicinity of London in *The Universal British Directory* in 1790. Further west Jewin Street, Clerkenwell and Friday Street, Snowhill, Holborn, Newgate, Warwick Lane, the Strand, and Piccadilly all had listings of salesmen operating on these lesser and greater thoroughfares, although the vicinity of Smithfield had by far the greatest number of dealers at nearly twenty. Until the formal establishment of an Old Clothes Exchange in 1843, the courts, alley-ways, closes, and pavements of these congested regions around Houndsditch and Petticoat Lane, Rosemary Lane, and Smithfield continued to receive the collected merchandise and witness the scrambling barters and sales among traders, before the piles of goods were redistributed to retail salesmen, both in London and the provinces.[78]

Among those who came to London to supplement their stocks of used apparel were clothes-brokers or salesmen from provincial towns and cities.[79] Unlike their London confederates operating shops and stalls in the metropolis, they did not have access to the wholesale supplies of used garments available in London. Thus, some with an

---

[77] Ibid. 437, 622, 757.   [78] Mayhew, ii, 26.
[79] Overseas traders came as well to deal in used clothing. This area also requires further research in order to determine the nature of the international trade.

expanding trade would look to the London second-hand market to supply those goods in greatest demand. Whether provincial dealers in second-hand clothes relied on local supplies or obtained stock from London, they ensured that used apparel could be bought throughout Britain. Shopkeepers designated as dealers in old clothes, salesmen, clothes-brokers, slop-sellers, and old-clothes men could be found in ports, industrial centres, and market-towns large and small throughout Britain.

Isaac Cotterell compiled a manuscript recording the major tradesmen of Bristol and, although it was never published, Cotterell's 'List of all the Merchants and Tradesmen in Bristol in October 1768' does reveal the presence of four 'sales shops' in the precincts of this port, on Corn Street, Wine Street, Castle, and on the 'Key'.[80] *Sketchley's Bristol Directory of 1775* contains evidence of a great many more men and women employed in the retail and wholesale trade in old or used clothing. Approximately eighteen tradesmen are noted in this directory, and many of these combine some role in the new retail-clothing trade, like tailor or shoemaker, with that of clothes-broker. Others, like Dennis Sullivan on the quay and William Andrews on Castle Street, are simply listed as salesmen, while Mary Ward is described as owning an old-clothes shop on St James's-back. The wholesale trade centred in Bristol is also revealed in the 'Ragg Warehouse' operated by Robert Arsley on Gloucester Lane.[81] Many hundreds of smaller, less affluent middlemen carried on some part of the trade in second-hand clothes, leaving little evidence of their affairs, being too humble to attract the notice of compilers of directories. Susannah Somers of Biggleswade, Bedfordshire, was one such dealer. Evidence of her small trade in second-hand goods was uncovered recently in the Bedfordshire Record Office exemplifying trades of this sort. An inventory of her belongings was made in 1770. At her death she had £17. 10s. 6d. in cash and approximately £40's worth of used apparel to offer to her customers, including women's, boy's, and girl's wear; hats, bonnets, shoes, petticoats, gowns of linen and cotton and the like. Its seems probable that Susannah did not restrict her trade exclusively to clothing. Her front room boasted a surprising assortment of earthenware—for example, 'Seventeen Earthen plates of different Sizes and

---

[80] B 21353, Bristol Reference Library. I am indebted to Sarah Levitt, curator at the Bristol City Museum, for drawing my attention to this source.

[81] *Sketchley's Bristol Directory of 1775* (1775, repr. 1971), 2, 15, 23, 24, 27, 28, 32, 36, 44, 57, 77, 78, 85, 87, 93, 101, 107.

Sorts & Do [ditto] Soup Dishes', as well as three earthenware teapots. Stock-in-trade was not distinguished from personal possessions; nevertheless, the numbers of some items identify them as likely trade-goods. Susannah would have exchanged or sold this stock, using the merchandise or money obtained for another round of barter or sale at her town market, seasonal fairs, or simply through local trade. Susannah earned what appears to have been a modest prosperity through this trade, as witnessed in the furnishings in bedroom and front room, yet there was no official record of her enterprise in either national or local directories of the county.[82] Small dealers like Susannah Somers, as well as the many nameless who bought and sold at local markets and fairs, worked in vital ways at providing the British public with items most in demand.

Aside from the unknown numbers of anonymous dealers in second-hand clothes there were hundreds listed in the many directories of the late eighteenth century. Over 250 shopkeepers and traders were catalogued as dealers in second-hand clothes in the four-volume *Universal British Directory*; in addition, almost 330 pawnbrokers were listed in that and other contemporary directories.[83] This network fed demand even at the lowest levels of society. The manner and frequency with which transactions took place had not been guessed at by modern historians, where demand and consumption is almost always considered in terms of cash-payments—markets outside this sphere received little formal attention. Yet without this flexible mechanism for payment, demand in the largest segment of society would have been considerably reduced. The sale and exchange of used clothing appears as an intermediary trade characteristic of a society in the throes of expanding production, wherein volume and variety are increasing, but productive techniques do not yet allow prices to fall to the level that permits generalized access to new goods. As a result of this flourishing commerce patterns of buying were altered, the poorer segments of the population could become accustomed to more frequent buying and selling as a consequence of this trade. The challenge for the cotton industry was to manufacture greater numbers of inexpensive, even cheap textiles, to tap the second tier of demand, to bring the majority of British society into the interplay of production

[82] Bedford Record Office. I am indebted to Miss Anne Buck for providing me with a copy of this inventory, which she uncovered in the course of her research.
[83] *The Universal British Directory*, vols. i–iv; *Bailey's Western and Midland Directory* (1784); *Bailey's Northern Directory* (1781).

and consumption that would come to characterize industrial Britain. The second-hand trade reveals a hitherto unrecognized depth to the demand rooted in pre-industrial and early industrial British society, explaining in part the manner in which consumer trades like the cotton industry developed in the eighteenth-century domestic market.

3
'WHO WILL BUY?': THE CONSUMER PROCESS AND THE COTTON INDUSTRY

THE importance of a study of consumption, as opposed solely to an appraisal of demand, rests on the fact that the former is the active response to the products themselves by the growing labouring and middle classes. Changes in choice and selection over time reflect the coming together of innovations in production and the needs perceived by collectives of individuals. To give a history of the evolution of consumption is not an easy undertaking even when the study focuses on one facet of consumption in the daily lives of the population. Nevertheless, it is vital to try to wrest anonymity from the consumers who bought the new sorts of textiles and who constituted the domestic market for the cotton industry. It is not possible to build precise representative inventories from each segment of society; on the other hand, the records that remain offer more than vague, impressionistic evidence about the cottons worn. Who wore cottons and what they chose to wear can be determined, as can the rate of proliferation of fabrics in conjunction with domestic consumption. Gauging alterations in mass consumption necessitates the use of indirect as well as direct methods; examining the additions of new cottons sold and the developments in cotton production, along with an examination of the evolving patterns of ownership of clothing and household furnishings.

EARLY PRODUCTS OF THE BRITISH COTTON INDUSTRY

The 1721 ban on almost all cottons officially eliminated the small amounts of pure-cotton textiles being manufactured in Britain. In 1720 a petition dispatched to the House of Commons from the boroughs of Weymouth and Melcombe Regis in Dorset on behalf of 'Merchants, Masters of Ships, Master Workmen, Weavers & Spinners of Cotton Wool' begged that the House exempt domestically manufactured cotton cloth from the prohibition, especially as in that area of the West Country it employed 'a great Number of poor People, who are incapable of other Work'. They hoped for encouragement from Westminster and requested that cloth of 'Cotton Wool ... be subjected to no other Duties than the Manufacture of British Linen is subject to'.[1] Their appeal was rejected. Thereafter, mixed fabrics of linen warp and cotton weft became the cornerstone of the early manufacturing industry, progressing to pure cottons later in the century. After 1721 the initiative passed to the Lancashire region, where comparable fabrics were already being woven. During the subsequent decades manufacturers of Manchester wares relied on the home market buying ever-greater stocks of fustians, tapes, checks, and other mixed textiles. It was on the success of these goods that the industry depended.

The trade records of Samuel Finney and Ralph Davenport reveal precisely the sorts of goods mentioned above as their principal stock-in-trade. Based in Manchester and in Truro, with warehousing in Bristol, both Finney and Davenport undertook excursions into the West Country, serving their customers *en route*.[2] Seven types of fabric described as 'Cotton Check' were commonly sold to their retail customers; as too were at least eight sorts of striped fustian, a fustian described as white 'Tufts', approximately seven kinds of pillow, assortments of 'Cotton' cloth, four species of white jean, five kinds of yellow canvas, several sorts of holland and linen checks, and over ten varieties of tapes, incles, and twist.[3] Davenport, Finney, and their confederates in trade sold merchandise whose characteristics were those of popular textiles. Most goods were hard-wearing and all the cotton fabrics were easily washable—qualities to recommend them to British consumers. The wealthy could continue to purchase Indian muslins for their wardrobe; however, the less-affluent British consumer looked in part to the printed linens and British-home cotton-linen chintz, checked cotton and linens, and other inexpensive patterned fabrics.[4] These goods were not made to the quality or in the varieties of the Indian fabrics, either in construction or in the range of colours available. However, prohibition and reordered the domestic market.

[1] *Journals of the House of Commons*, 7 Geo. I, 1720, 418.
[2] Chapter 4 contains a more extensive account of the trading practices of these two men and the geographic and commercial scope of their trade.
[3] DFF/20/17; DFF/20/41–51, Cheshire Record Office, Chester.
[4] The ban on East Indian calicoes specifically excluded four sets of textile products—all-blue calico, muslins, neckcloths, and fustians were permitted under this legislation. 7 Geo. I, c. 7.

The legislative barrier obliged consumers to try to make do with the improving British products.

A few manufacturers had been making chintzes since the beginning of the eighteenth century in imitation of the Indian materials. The erection of legal impediments after 1721 stimulated the development of a greater assortment of textiles comprised in some part of cotton yarn. Fustians, fabrics of mixed cotton and linen composition, had been woven for centuries, as had fabrics of wool and cotton, and silk and cotton. The popularity of these multi-fibre materials had spread across Europe from the fourteenth century. Only the weavers' skills limited the variations that could be produced. What was crucial for the cotton industry was that this tradition of invention and innovation continued in the production of cotton-based fabrics, and that these were being systematically produced to fill the void that now existed.

John Holker's report on the Lancashire textile industry in the 1750's provides a unique sampling of the varieties of textiles then produced.[5] A comprehensive price analysis cannot be determined from this report, but the samples, supplemented by the brief descriptions of the swatches, give a clear indication of the products made by English artisans and sold widely in Britain. In all, the report listed 115 types of cloth. Of this total four were linen made in London, four were silk from Spitalfields, six silk-and-wool and all-wool materials were the product of Norwich manufacturers, and two light-weight wools came from Yorkshire. Thus of the total, ninety-nine of the fabrics included in the report originated in Lancashire and less than eighty could be described as cotton. Table 3.1 sets out the assorted fabrics as Holker recorded them.

Checks comprised the largest single group of fabrics: Holker included seventeen assorted samples of checked cloth.[6] Of these not all were examples of cotton or even cotton-linen fabrics. In fact only one of the check samples collected by Holker could be purchased in an all-cotton variety. Two of the other materials of red and navy-blue checks were made of linen and wool. The presence of only one all-cotton check, as well as several sorts of linen-wool checks, reflects the technological deficiencies that hampered production at this time.

---

[5] Holker was a failed Jacobite, employed by the French government to obtain samples and information about the flourishing cotton industry, and this he attempted to do at great personal risk during several trips to Britain in the 1750s. G.G. 2, Livres Eschantillon . . ., Musée Centrale des Arts Décoratif, Paris.

[6] G.G. 2, nos. 1–17, Musée Centrale des Arts Décoratifs.

TABLE 3.1 *Lancashire cotton textiles in Holker's report, c.1750*

Cotton-linen		Cotton	
Check	17	Check	1
Chintz	13	Nankin	1
Cotton Diaper	6	Velvet	7
Cotton Stripe	10		
Fustian	11		
Grandurelle	1		
Handkerchief	1		
Herringbone	1		
Jean	4		
Striped Dimity	3		
TOTAL	67	TOTAL	9

Without a means of producing cheap and abundant stocks of strong cotton warp, manufacturers were obliged to use linen or Indian cotton warp; while inadequacies in the techniques of dyeing a colour-fast red made wool yarn preferable to cotton in patterns where red predominated.[7] One can only speculate as to whether these technological deficiencies measurably affected the levels of consumption.

Much of the diversity of this collection of checks lay in the many combinations of colours used, including navy blue, light blue, red, white, and yellow, as well as light and dark green. Of the seventeen, nine were intended for clothing and eight for furnishings, with the cost of these materials running from approximately twelve to seventeenth pence per yard. As each of the check fabrics listed was accompanied by a sample of cloth, other differentiating features could also be discerned. The varying quality in the yarn and the construction of the fabrics was very apparent; the cheaper sorts were comprised of coarser fibres and the weaving too was less fine. Options increased with a few added pence per yard, and one could select a slightly finer and more

---

[7] After lengthy, expensive, and time-consuming efforts to discover the secret of colour-fast red dyeing, manufacturers finally had the mystery revealed to them by a Frenchman, Louis Borelle. He offered to explain the French method of dyeing to the Manchester Committee of Trade in 1784 and for his trouble was awarded £2,500, once tests proved that this did indeed work—such was the importance ascribed to mastering this colour for the cotton industry. A. P. Wadsworth and Julia de Lacy Mann, *The Cotton Trade and Industrial Lancashire* (1931), 179–81.

densely woven cloth to make into shifts, aprons, or shirts. Furniture checks, on the other hand, were routinely heavier and more durable than those produced for clothing, and once again this differentiation was plainly visible in the samples. The final variations in the checks were based on the size of the checks and fine lines that made up the patterns. Clothing checks tended to be small or medium-sized, while those for furnishings were bolder, larger, and brightly patterned. Sample number nine, for example, was a very fine small check; the fabric itself of cotton and linen, although Holker relates that an all-cotton variety of this check was also available. Number fourteen was a yellow-and-red check, the very wide yellow blocks outlined in red, intended for use as curtains and chair-covers. The possible combinations and permutations of colours, check, and line seem endless. It may be, however, that production was restricted to the approximate assortment included in the 1750 report, for although the demand for novelty would have pressed manufacturers for continuing inventiveness, technological and organizational inadequacies in the mid-century seem to have engendered limitations irksome to merchants and customers alike.

The Bristol merchants Peach & Pierce told one customer of the difficulty in obtaining checks outside the range offered for his inspection. In the spring of 1753 Peach wrote:

I have been as exact to patterns as the times will allow[.] Our Manufacturers are under no Command and 'tis impossible to get them to make all small and dark Patterns without extra-ordinary price out for the future I will preserve those patterns by me on purpose ... for they are saleable only with You.[8]

In this case the customer's insistence on a dark-blue, small-checked cloth added substantially to the cost of the fabric, and Peach & Pierce appeared reluctant to stock these more costly goods. Two years later the question arose again and the merchants wrote that, 'I am sorry my Checks have made it impossible to get a large Quantity of all dark Pattern'.[9] Clearly the breadth of consumer demand was not met by the existing stock. The productive capacity of the cotton industry and its affiliated printers and finishers could not yet guarantee a quick response to customer requests for non-standard textiles. Some orders which diverged from the accustomed stocks met with delays and added expense, all of which acted to impede the satisfaction of consumers.

[8] Philip L. White (ed.), *The Beekman Mercantile Papers* (1956), ii, 536.
[9] Ibid. 539.

Yet even with such impediments the market for checks waxed larger with the passing years.

One of the Lancashire products which acquired unexpected success in Holker's period was the furniture-checks. 'Checker'd' textiles had found part of their market in 'Bed-quilts' before the end of the seventeenth century.[10] Of course at that time the most popular soft furnishings came from India; but with that competitor considerably hampered after 1721, the substitution of brightly patterned English furniture-checks proceeded apace. Greater attention went into the creation of a new range of furniture-checks and thereafter these bright, washable textiles found considerable success as coverings, wall-hangings, curtains, and draperies. Samuel Touchet testified before a committee of the House of Commons in 1751 that 'the Manchester Trade has continued to increase still within these Two or Three Years'. Moreover, Touchet attributed this growth to 'a particular Fashion in Furniture'[11] that had stimulated consumption. Chronic, intermittent disruptions in the export market made the home market all-the-more attractive to the manufacturers of cotton textiles. Witness the speed with which window curtains were deemed necessities and adopted by a significant segment of the population from the late seventeenth century onwards, as revealed by Lorna Weatherill's sampling of inventories. Ralph Davenport followed this trend. In 1731, for example, when the merchant set up a household in Truro, his list of household furnishings very naturally included 'Curtains and Rods'.[12] The 'particular Fashion in Furniture' noted by Touchet two decades later had taken strong hold and assured the extensive consumption of British checks by establishing yet another commodity as essential to gentility or even respectability; as a necessity, not a luxury. Window curtains were visible signals, fluttering flags of relative affluence that marked the windows of the genteel and the aspirants to that standing, in city, town, or village.

In 1753 the orders sent by Peach & Pierce to James Beekman listed thirteen assorted qualities of linen check, of four main widths. In the same year only two sorts of cotton checks were shipped, and these were provided in two widths only. Within the next ten years more and more cotton checks were added to their inventory. By 1770 the

---

[10] J. F., *The Merchant's Ware-house laid open: Or, The Plain-Dealing Linen Draper* (1696), 6.
[11] *Reports relating to chequered and striped Linens* (1751), *Reports from Committees of the House of Commons* (1737–65), ii, 293.   [12] DFF/20/36, Cheshire RO.

number of qualities and widths of cotton checks dispatched by Peach and Pierce equalled that of the linen checks—thirteen. This rate of proliferation was commonplace for many varieties of textiles. In 1767, for example, the invoices revealed five sorts of pillow, an increase of two over the three sorts listed in 1750, while in 1771, seven sorts of pillow were available. Another example of diversification appears in a series of Bristol inventories, wherein ten qualities of thickset were being sold as early as 1757 through Peach & Pierce.[13] Choice did not await the introduction of mechanization; variety and selection were forthcoming from the domestic system, with consumer demand a spur to further expansion of production.

Printed chintzes made up the next largest sort of fabric collected by Holker, who listed thirteen sorts.[14] These were the most expensive light clothing fabrics, costing from about three shillings and sixpence to four shillings per yard. The mixed cotton-linen fabrics were of a uniform length and width, and here too the variation in the textile depended on the patterns printed on the cloth or on the special finish. In common with checks, there was no evidence of dramatic variations in the quality of the woven cloth: in fact, this cloth appeared to conform even more stringently to a single standard. Variety was found in the printed patterns. A satin-like finish worked on a printed chintz raised it 15 per cent in price above the other chintzes, which in turn were almost four times the cost of the checked materials. But though the extra cost of the finish made this fabric dearer than the regular chintz, the customer was provided with an even greater verisimilitude between the British-made chintzes and the brocades and printed silks which it sought to emulate.[15]

The Manchester and Blackburn manufacturers depended heavily on the expertise of the London printers of novelty in the finished products, the fabric being a medium for the printers' arts. This

[13] Invoices, Robert & Nathan Hyde, 29 Mar. 1757; 17 Mar. 1767; 20 Mar. 1771. Beekman Papers, New York Historical Society, New York.

[14] G.G. 2, nos. 88–100, Musée Centrale des Arts Décoratifs.

[15] Godfrey Smith explained in his 1756 book the origins of the botanic prints which had such a wide appeal. 'With respect to the drawing of the patterns for the calico-printers, they are, for the generality, in imitation of the flowered silk-manufactory, with such variations as may best answer the nature of the different sorts of works, of which there is a great variety. The principal are the chintzes, in which they imitate the richest silk brocades, with a great variety of beautiful colours ... The fashion, as with the brocaded silks, has run upon natural flowers, stalks and leaves ... or in sprigs and branches carelessly flung, in a natural and agreeable manner.' Quoted in Florence M. Montgomery, *Printed Textiles* (1970), 24.

dependence continued to the second half of the century when the regions of the north-west generated printers of their own. But whether in London or in Lancashire, the pattern-drawers and printers expended every effort to describe attractive patterns on paper and prepare these for the waiting cloth. The requirements of the pattern-drawer were many, as Campbell explained. At the least the trade demanded 'a fruitful Fancy, to invent new Whims to please the changeable Foible of the Ladies, for whose Use their Work is chiefly intended'. But Campbell went on to assure the parents of sons blessed with only modest talents that, 'It requires no great Taste in Painting, nor the Principles of Drawing; but a wild kind of Imagination, to adorn their Work with a sort of regular Confusion, fit to attract the Eye but not to please the Judgement'.[16] Once the patterns had been designed, printers took over the task of applying the dyes according to pattern. Having worked for generations on East India cottons, the textile printers applied themselves to the British manufactures and assisted considerably in establishing the popularity of these fabrics. Campbell could not claim for the English printers a pre-eminent expertise in the trade by 1750, particularly when compared to the East Indian printers and the products of their craft. 'Ours come short of theirs both in their Beauty, Life and Durableness', Campbell temporized. None the less, in the largely protected market within Britain the artisans were able to improve their techniques and expand their sales to such a degree that, as Campbell averred, 'the whole Kingdom is furnished with Commodities of this sort'.[17]

Cotton hollands and cherydery were in the same general category of light clothing textiles as were chintzes. The first was made of cotton weft and linen warp and usually decorated with stripes, both thin and broad, in colours similar to those of the checks. The cotton hollands were woven in two widths and in several qualities, thus adding to the numbers of this textile available to the public. These ten varieties of cloth formed a middle quality of goods in Holker's collection between the inexpensive checks and the more costly chintzes.[18] Holker also listed eight sorts of cherydery in his report, all of silk and cotton in composition; attractive light fabrics based on an Indian precursor.[19]

[16] R. Campbell, *The London Tradesman . . . All the Trades, Professions, Arts, both Liberal and Mechanic, now practised in the Cities of London and Westminster* (1747), 115.
[17] Ibid. 115, 116.
[18] G.G. 2, nos. 21–30, Musée Centrale des Arts Décoratifs.
[19] Ibid. 33–40.

A Particular of Goods to be expoſed to Sale by the Eaſt-India Company, in September, 1676. Viz.

Pieces.	
23890	Long Clothes White
9468	Ditto Brown
3200	Ditto Blew
46954	Salampores White
3835	Ditto Brown
16139	Percallaes
15080	Mores
9326	Ginghams
5515	Ditto Coloured
3550	Izzarees
1320	Diapers
2473	Betteleis White
107	Ditto Brown
8280	Ditto Oringil
640	Neck-clothes
5000	Allejaes
133	Sheets
8382	Dungarees
28833	Sannoes
1236	Humhums
2902	Cossaes
1123	Mulmulls
5700	Tassities
2430	Ditto Raw
6593	Nillaes
920	Quilts large
400	Ditto ſmall
1440	Chints largeEruuch
6500	Ditto Broad
7400	Ditto Narrow
3200	Ditto Serunge
7680	Ditto Caddy
5809	Tapietſes broad
5010	Ditto Narrow
4435	Niccanees
9060	Guinea Stuffs
9300	Bravels
12592	Pautkaes White
9070	Ditto Brown
3400	Ditto Blew
982	Dungares White
81	Ditto Brown
570	Derebands Small
	Brauauls
35305	Baffs Broad White
9120	Ditto Blew
1755	Ditto Narrow White
37900	Ditto Narrow White
1005	Ditto Brown
43	Parcallaes
561	Geetings
43	Mulſers
	Raw Silk Bengale — Bales 108
	Raw Silk China — 24
	Stitching Silk Dyed — 2
	Florctta Yarn — 40
1601	Cordivant skins
	Musk in Cod — Potts 3
	Pepper Black — Ba.7886
	Ditto White — 144
	Ditto Damaged and Duſt — 19
	Cotton Yarn — 352
	Carmenia Wooll — 318
	Coffee — 106
	Cardamoms — 209
	Olibanum — 143
	Seedlack — 4
	Sticklack — 161
	Tumrrick — 363
	Rice — 320
	Cowrees — 500
	Saltpetre — 408
	Ditto Fine — 108
	China Roots — Canniſters 78
	Indigo Lahore — Barrels 275
	Ditto Cirques flat — 106
	Ditto round — 3
	Indigo Duſt — 37
	Benjamin — Cheſts 42
	Lapis Tutia — 13
	Salarmoniack — 51
	Green Ginger — Jarrs 579
	Tincal — Dippers 184
	Sapan Wood — Sticks 2
	Sandal Wood — C
	Red Earth
100	Buffloe Hides
631	Indigo Skins
611	Glews

FIG. 1. The range and variety of textiles imported from the East Indies is reflected in this auction handbill.

FIG. 2. English calico printers, using the block printing technique, produced approximately one million yards of printed calico in 1711.

THE
CASE
OF THE
Weavers of the City of London and Parts Adjacent,

Humbly represented to the Honourable the House of Commons,

Sheweth,

THAT about the latter End of the Reign of the late King *William*, the Weavers in and about *London* were under the same Discouragement and Distress as they are now, occasion'd by the great Importation of *East India* Wrought Silks, Printed Callicoes, &c. That the Manufactures of Wool and Silk were then under such a visible Decay, and in such imminent Danger of Ruin, that the Manufacturers were forc'd, *just as they are now*, to lay their Condition before the Parliament, in the humblest Manner, imploring such Relief as should seem reasonable to their Honours in such a deplorable Case.

THE rest of the Manufacturers in Wool and Silk, in many, if not in most or all the Parts of the Kingdom, sensible of the like Decay in their Trade, and of the Want of Employment for their Poor, and sensible of its being owing to the same Cause, namely, the Importation of so great a Quantity of Printed and Painted Callicoes and Wrought Silks from Abroad, join'd at the same Time with the said Weavers, and represented their Case likewise to the Parliament.

THE Parliament having fully examin'd and consider'd the said Representations, were so satisfy'd of the Reality of the Grievance, *as set forth in the said Petitions respectively*; as also of the said Wearing and Use of Painted Callicoes and Wrought Silks being the Cause of the same: That an Act was passed in the Year 1700, forbidding the Use and Wearing of them on any Account whatsoever.

THIS Act for some Time prov'd of so great Advantage to the Nation, that it is evident, that for several Years after the passing thereof, the Trade reviv'd; the Manufactures of Wool and Silk were restor'd, and flourish'd, to the very great Improvement of the Manufactures themselves, to the Enriching the Nation, and to the evident Advantage and Employment of the Poor.

BUT the clandestine Traders, after some Time, finding Ways to run in upon us great Quantities of *East India* Goods, that Act appear'd insufficient, totally to suppress the Wearing and Use of the said Printed Callicoes and Silks.

ALSO our own People were hereby encourag'd to fall into, and improve very much, the Art of Printing and Painting Callicoes and Linnens at Home: Which said Art was very little known or practis'd in *England* before that Time, tho' it has been fatally increas'd since.

THE Humour of Wearing the said Callicoes, &c. thus returning, and prevailing among the People to such a Degree, and with such ill Consequences, as in the Petition is set forth; the Distress of the Weavers, and the Decay of their Trade, is naturally return'd with it, as the Effect following the Cause.

THE Goods made by the Masters lye on Hand unsold, the Market for them being supply'd by the Sale of Printed Callicoes and Linnens, as above.

SEVERAL Sorts of Worsted Stuffs, and Stuffs made of Silk and Worsted, are quite lost and out of Wear, being wholly supply'd by the said Callicoes and Linnen.

THE Poor are left unemploy'd, and are either reduc'd to inexpressible Misery and Distress, or forc'd to fly to foreign Countries for Work, leaving their Families to perish, or to be supported by the Alms of the Parish where they dwell.

Therefore they humbly hope, the same Reasons moving this Honourable House, they may obtain the same Relief; and that the Use and Wearing the said Callicoes and Linnens may be more effectually prohibited and prevented.

All which is Humbly Submitted, &c.

FIG. 3. Hundreds of pamphlets, petitions and tracts were distributed in 1719–20, demanding the prohibition of printed East-Indian textiles from England.

FIG. 4. Gowns made of printed Indian cotton continued to be in demand even after the 1721 ban.

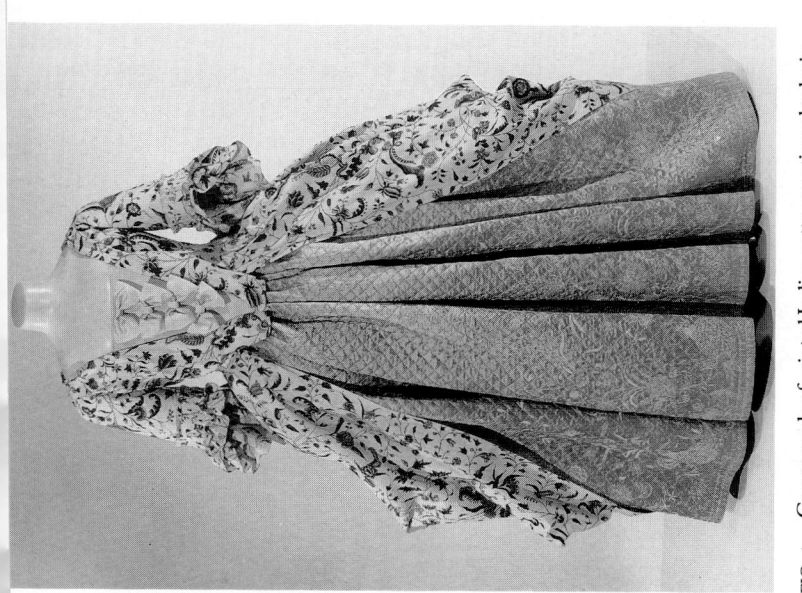

FIG. 5. Dressing in a gown made of a prohibited fabric gave some wearers an extra fillip of pleasure, defying the law in the interest of fashion.

FIG. 11. Many women maintained a small trade in used apparel, serving their neighbourhoods and local communities.

BROWN,
TAYLOR and HABIT-MAKER,
(No. 77.)

Broad Street, near the Royal Exchange.

BEGS Leave to lay the following Estimates to the Confideration of the Public, when at least 2½ per Cent. may be faved, and the Cloth and Workmanfhip, as Good as any of the Trade in LONDON.

	£. s. d.		£. s. d.
A Frock Suit of Superfine with Silk Knee Garters	4 10 0	A Pair of Four Thread Superfine Knit Stocking or Corduroy	0 17 6
A Half Dreft Suit	5 5 0	A Pair of Genton Velvet ditto	1 14 0
A French Frock Suit ditto	4 16 0	A Superfine Kerfey Beaver Surtout Coat	1 13 0
A Full trimm'd Suit ditto	5 10 0	A Superfine Bath Beaver	1 6 0
A Frock Coat and Waiftcoat	3 13 6	A Frock Suit of Livery all Cloth	3 5 0
A Frock Coat Ditto	2 12 6	A French Frock Suit of ditto	3 10 6
A Pair of Breeches ditto	2 18 0	Shag Breeches extra	0 7 0
A Frock Suit of Second	—	Ditto laced with fhag Breeches	4 10 0
A Coat of Ditto	3 13 6	A Thickfet Frock and Waiftcoat	2 10 0
A Waiftcoat of ditto	1 10 0	A Fuftian Frock and Waiftcoat	2 5 0
A Pair of Breeches	0 15 0	A Surtout Coat for ditto	2 0 0
A Pair of Silk Knit	1 13 0	A Couch Box Coat lined through	3 5 0
Ditto the Beft	1 18 0		

Ladies Riding-Habits at the following Prices:

	£. s. d.		£. s. d.
A Fine Kerferymere Riding Habit	5 0 0	A Fine Jean of Jennet Ditto	3 3 0
A Ladies Cloth Ditto	4 14 6	A Nankeen Ditto	3 0 0

Trimmed and any other Sorts of Habits equally Reafonable.

Regimentals and Uniforms in the neweft Manner, Alfo Gentlemen fitted for Abroad, Merchants and Captains with any Kind of Cloathing for Exportation on the beft Terms. All Gentlemen who are above a middle Size, muft allow in Proportion, and the Difference in Blue, Grey, and all Colours in Grain, gilt Buttons, Velvet Collars, and Waiftcoat Sleeves muft be paid for extra.

Gentlemen having Cloaths to difpofe of, from Mif-fitting or Alteration in Fafhion, he takes thofe back, allowing more than in general is allowed, being able fo to do from a fuperior Demand he has for that Commodity. Contracts to ferve Gentlemen with Cloth by the Year, at the following Prices, and takes them back, each Suit to be returned, and an equal Proportion of Money paid on the Delivery of each Suit. For four Frock Suits fuperfine £15. Three Ditto £11 10. Two Ditto £7 17 6. And each Gentleman is at Liberty to wear gilt Buttons and Silk Garters, but Grain Colours and Waiftcoat Sleeves muft be paid for Ext.a.

N. B. I deal for READY MONEY only.

Printed by THOMAS PARKER, No. 4, Bull Head Court, Jewn Street.

FIG. 10. Brown described to potential customers his trade-in policy, which would stimulate the sale of new clothing.

FIG. 12. Jewish clothes dealers figured prominently among the urban street dealers, many of whom began their modest trade as new immigrants to Britain.

FIG. 13. The wielder of this mop sports a number of stylish accessories from buckled shoes, to white stockings and neat cap.

FIG. 14. The Strawberry seller, shown here wearing a print gown, exemplifies the class of consumer to whom British cotton-manufacturers directed many of their products.

The cheryderies, as well as the hollands, were used in women's clothing, and the cost of the silk and cotton fabrics began at a third more in price than the striped hollands. The most expensive cherydery cost the same as did the printed chintz. Thus with the cost of the cherydery ranging from 2 to 3 shillings per yard the price of fabric for a gown could cost as much as 15 shillings. However, it is important to remember that this price was the upper end of the scale and even at that price was well within the means of the middling sector of society, adding to the variety of materials available to the discerning consumer.

Besides the many examples of dress-goods, Holker collected samples of the ubiquitous handkerchief. The principal emphasis in Lancashire appears to have been on the production of linen handkerchiefs in patterns of blue, red, and white checks, as well as white dots resist-printed on red or blue backgrounds. Holker included several examples of these sorts. Another was made of white linen, with a border of red cotton. An attractive blue, red, and white checked handkerchief, produced in and around Manchester, cost approximately 10½ pence and was reported to have an extensive market and wide popularity in England.[20] Manchester also provided the only example of a cotton-linen handkerchief, and in order to secure a colour-fast red in the red, white, and blue checks the manufacturers depended on Indian red yarn. The last of these Lancashire samples was a facsimile of the East India style of handkerchiefs, while the remainder were silk produced in London and its environs. At this time the cotton manufacturers produced few kinds of cotton or cotton-blend handkerchiefs, but those that were made had an extensive market share, epitomizing the products characteristic of this trade, being both attractive and inexpensive—an ideal accessory for labouring women and men.

Manchester's renowned all-cotton velvets were the largest category of goods produced from purely cotton yarn at this time. While these fabrics found a wide market in Britain and abroad,[21] there were in fact only seven sorts of velvets presented in the Holker sample book, plus one example of a silk-and-cotton velvet. It is doubtful that this was a complete range of the cotton velvets. More than likely these were all Holker could obtain.[22] Production techniques of the plain and

---

[20] Ibid. 77–9, 84.

[21] Wadsworth and Mann, 174–5, describes the success of this fabric.

[22] Wadsworth and Mann relate Holker's services to the French government as an industrial spy, after his escape from prison where he was incarcerated as a Jacobite. He acknowledged in his report that samples and information about the manufacture of certain goods was difficult to come by.

flowered velvets were jealously guarded, for these fabrics represented a significant breakthrough in the manufacture of an all-cotton cloth. Even in 1755, when Holker had established a rival French manufactory in this product, he acknowledged that the superior skill of the Lancashire craftsmen produced a full-surfaced, heavy-cotton velvet at a cost lower than any comparable French product.[23] These velvets were the most expensive textiles in the collection, running from about 8 shillings per yard to over 12 shillings. The silk-and-cotton velvets were a quarter as much again.[24] From these goods the British consumer obtained a brightly patterned or plain, high-quality garment that was also warm and hard-wearing. The initial outlay for velvets of this sort would be well repaid in the years of wear and the exemplary appearance, for Holker testified that the look of the velvets improved with wear, while at the same time being extraordinarily durable. An investment in this sort of British-made cotton velvet would provide value not only for the wearer in the quality of his or her clothing, but also in the assured resale value, enabling the buyer to purchase a newer coloured or printed velvet.

Lancashire manufactured fustians in the greatest varieties. Samples of these fabrics were scattered throughout the report and can be divided into several categories. The first assortment was the heavy, un-napped cloths such as jean, nankin, ticking, hooping, grandurelle, and herringbone. The samples collected by Holker of ticking, hooping, and two sorts of jean were made entirely of linen; the nankin was an all-cotton cloth; while four sorts of jean, one of grandurelle and herringbone, were cotton-linen mixtures. These fabrics constituted a humble, but indispensible part of the dress of the ordinary man or woman in the making of footwear, stays, hoops, jerkins, pockets, and breeches, among other things. The second sort of fustians were those with raised woven patterns like the three kinds of self-corded dimity and the six varieties of diaper decorated with small, medium, and large raised floral designs. This last group was made entirely of cotton weft and linen warp. Finally, the report included samples of napped and corded cloths, like pillow, mentioned above as a staple at the turn of the century, as well as barragon and thickset. Holker had three samples of the former and two each of the latter fabrics included among his selection. These examples were by no means exhaustive; Holker noted that the barragon was also sold in an all-cotton variety,

[23] Wadsworth and Mann, 201–2.
[24] G.G. 2, nos. 54–60, 72, Musée Centrale des Arts Décoratifs.

although he did not have a swatch for examination. The prices ranged from the very cheapest hooping at 12½ pence per yard to the higher priced diapers, jeans, and thicksets, which could cost from 18 pence per yard to four shillings per yard or more. Holker advised his sponsors that this group of twenty-five fustians probably accounted for the largest part of the trade in cotton textiles within Britain, and occupied the greatest portion of the manufacturing capacity of Lancashire.[25]

Throughout the 1750s and 1760s production in the cotton trade was directed towards diversification, creating additional choice for consumers who were growing more convinced of the desirability of these products. Just as cotton textiles appeared in greater numbers, so too did the many assorted accessories: buttons, buckles, caps, handkerchiefs, hose, and mittens, that multiplied over the first half of the eighteenth century, reflected a strong and persistent consumer demand. All these articles fell within a broadly middling price-range, running from the least expensive buttons and hose to the more elaborate, better quality articles. One 1758 invoice exemplifies the proliferation of goods for the insatiable consumers, in both dress accessories and accoutrements. In it were listed eight types of shirt-buttons and fourteen sorts of gilt, plated, tin, and horn buttons designed for various parts of the clothing. The invoice also included ten kinds of garters, 'check', scarlet, 'spotted', and 'white Lettered', to name a few; four qualities of women's mitts; two colours of muffatees; thirty-two sorts of hose; seven types of buckles; and six kinds of two-seam caps.[26] Furthermore, this selection was not unusual. A 1757 invoice listed an equally dizzying assortment of hose, including cotton-thread hose, cotton-and-worsted mixed hose, as well as patterns described as marbled, 'Harlequin', 'Mosaick', and 'New flash' men's hose, plus 'blue and white Clocks' women's hose. In all, over forty sorts of men's, boy's, and women's hose were listed.[27] The popularization of assorted types of stockings was another of the successful innovations of the period. This sort of pure cotton and mixed cotton-and-worsted hose gained customer allegiance across British society, as cotton stockings were inexpensive, attractive, and easy to care for. The middle and upper ranks could buy cotton hose as an alternative to the more costly silk variety, but for the working classes cotton stockings

---

[25] Ibid. nos. 18–20, 31–2, 42–52, 61–71.
[26] *The Beekman Mercantile Papers*, iii, 1, 404–9.
[27] Invoice, Peach & Pierce, 29 Mar. 1757, Beekman Papers.

conferred an apparent fashionability. The development and diversification of this trade centred on Nottinghamshire and Derbyshire, and has been extensively documented as the principal product of the midlands cotton industry. Much of the industrial and commercial prosperity of those areas could be attributed to the general consumption of endless varieties of cotton stockings manufactured in those regions.[28]

New articles appeared at regular intervals throughout the second half of the eighteenth century as contemporary events and fashions spurred the creation of novel buttons and handkerchiefs. 'Death head' buttons in four different colours were introduced in the 1760s. At the same time campaign buttons of various sizes and colours appear in the invoices.[29] Handkerchiefs were an essential part of the female dress during the whole of this period, and their low cost encouraged the production of an enormously varied range. In addition to the handkerchiefs made of new fabrics, there were also the unlimited contemporary topics of political or local interest, subjects suitable for printing on handkerchiefs. The 'Map of England' and a depiction of metropolitan trades called 'London Cryes' were included in orders of handkerchiefs.[30] It is difficult to believe that some enterprising printer would not also have turned out 'Wilkes and Liberty' handkerchiefs in addition to all the other paraphernalia manufactured during the heyday of the Wilkes campaign.[31] Thus the selection available to consumers expanded not only with the addition of cotton, but also with the patterns and prints on hand. By 1776 at least ten sorts of cotton-check handkerchiefs were being manufactured in Lancashire, as well as twenty-seven other kinds of handkerchief—such as printed cotton, black cotton, silk and cotton, muslin, silk and muslin, malabar, malabar blue, barcelona, romall and silk, and linen handkerchiefs.

The varieties of goods described by John Holker, and produced thereafter, were not simply inventories of interesting textiles. Rather

---

[28] S. D. Chapman, 'Enterprise and Innovation in the British Hosiery Industry', and *Textile History*, 5 (1974) *The Early Factory Masters*, (1967), 17–27, 145–6.

[29] *The Beekman Mercantile Papers*, ii. 837.   [30] Ibid. 946.

[31] As John Brewer points out, they produced everything else, from ceramics suitably decorated, prints, medals, pipes, pewter pots, cakes, wigs (with forty-five curls), and even a full radical suit. The predilection for commemorating the mighty champion, of whatever sort, on the humble ephemera of daily life was a bonanza for manufacturers, who could not seem to produce too much while the craze lasted. John Brewer, 'Commercialization and Politics', in Neil McKendrick *et al.*, *The Birth of a Consumer Society* (1982), 238–40.

they chronicle the relationship between the producers and consumer in Britain: the one impelled to produce an ever-wider choice of materials to meet demand and the other increasingly preoccupied at all levels with the purchase and display of new goods.

CONSUMPTION OF COTTON CLOTHING AND THE GROWTH OF THE COTTON INDUSTRY

The dress of men, women, and children changed over the course of the eighteenth century. Modifications and outright innovations in style occurred; however, the most dramatic alterations could be found in the progressive shift in the textile composition of dress. The proliferation of cotton accessories, as well as cotton textiles, testified to the strong demand and growing levels of domestic consumption in one of the most dynamic economic sectors of Europe. Documentation of the changing composition in popular clothing presents some difficulties. Fortunately an excellent source remains in legal records like the *Proceedings of the King's Commission of the Peace . . . for the City of London and . . . the County of Middlesex*, otherwise known as the *Old Bailey Records*. Through these statements the reader is admitted into the lives of an enormously varied cross-section of society: all those who were victims of crime or caught in a criminal act in the precincts of London and the adjoining regions. Houses, lodgings, and shops were routinely burgled for the clothing and textiles owned by the inhabitants. These items were easily turned into cash, either through a private sale or by pawning the proceeds at any one of the hundreds of pawnshops scattered across the metropolis. Prosecutions frequently contain enough information on the victims and the items stolen to enable a correlation to be made between the rank or occupation of the victims and the clothing and textiles which they owned. Legal records, supplemented by similar newspaper accounts, tradesmen's records, private papers, and inventories, provide extensive and continuous documentation of the use of cotton in dress throughout the eighteenth century.[32]

---

[32] In the course of my research into this topic as a doctoral student I discovered that legal records yielded much more than information on legal or criminal matters. The *Old Bailey Records*, and similar assizes material, have proven to be one of the most valuable sources in the systematic investigation of ownership of goods among the common people, one that had not been sufficiently recognized by social and economic historians. Court records provide documentation over the long-term that gives access to a cross-section of the population. These records help historians to trace the shifts and changes

Fustian was long known as the stuff of working-men. One case prosecuted at the Old Bailey in 1715 illustrates the status of some of those who routinely wore fustian. Cloth coats and a fustian frock were stolen, and from the court record, 'It appear'd that the 3 Persons who lost the Goods were Dust-men; and being at Work in the Street, hung their Cloaths upon the Pales, from whence the Prisoners took them'.[33] John Fowls, was likewise unable to resist the lure of 'a Fustian Frock and Waistcoat, 3 Shirts, and other Goods' which he absconded with from a glazier's shop where he had been working.[34] Cotton fabrics, whether of pure cotton or mixed composition, continued to be much in demand even after the legislative prohibition of 1721. The heavier cotton-linen sorts were useful additions to a working-man's wardrobe. Notices of runaway apprentices and other felonious youth contained a significant proportion described as wearing fustian frocks, dimity waistcoats, checked shirts, plus other heavy fabrics of this sort.[35] Fustian, dimity, and checked cotton-linen figured routinely in the dress of working boys and men, changing only in so far as new varieties allowed for greater choice later in the century.

It is difficult to determine how general the continuing use of pure cotton clothing was after the ban. Most owners of clothing suddenly declared to be proscribed may have persisted in wearing them, but with the constant threat of denunciation.[36] Cotton clothing was not as ubiquitous in the decades after 1721 as it became from the mid-century onwards. None the less, it was not unusual to find references in the court transcripts of thefts of '2 Cotton Shirts' the property of a sailor Benjamin Cook, or 'a Cotton Gown, 2 linen Aprons' from John

in common ownership of goods. John Styles has also begun using legal records in this manner, and from them seeks to discern the common dress of the eighteenth century, adding further the extant information on the social and economic history of dress.

[33] *Old Bailey Records*, Jan. 1714/1715, 5.

[34] Ibid., Dec. 1714. Many county-assizes records recount similar thefts of clothing, but without being specific as to the fabric composition of the garments.

[35] *The Stamford Mercury*, 4 Apr. 1728; *The Northampton Mercury*, 4 Feb. 1733/4, 23 Dec. 1734; *Old Bailey Records*, Oct. 1732, 234; William Le Hardy (ed.), *Buckinghamshire Calendar of Quarter Sessions Records* (1958), v, 41.

[36] *The Northampton Mercury*, 5 July 1736, notes a community effort to foil the prohibition against Indian fabrics. 'Last Sunday a great Disturbance happened at the new Church in Horsley Downs, occasion'd by some Informers who went thither to seek for prohibited India Silk and Callicoe Gowns; but being known, the whole Congregation were apprized thereof, and the Ladies who were under any Uneasiness at their Appearance, retir'd to the Vestry, where fresh Clothes being convey'd to them, they chang'd their Apparel, to the great Mortification of the Knights of the swearing Order, who were thereby defeated of their Golden Hopes'.

Cotterell's wife.[37] Penelope Coleman was indicted for stealing a pair of gold earrings, a pair of stays, and a calico apron from Daniel Garshier's young daughter; while Mary Robinson stole 'one Pair of Stays, one flowered Cotton Gown, a Leghorn Hat, a Flaxen Sheet and two Waistcoats' from John and Elizabeth Robinson, with whom she lived as a lodger.[38] Robert Leake was sent for in September of 1745 when a thief was found in his house coming down the stairs with two of his wife's cotton gowns under her cloak. His report of the interrogation revealed a rather odd motive for this larceny. 'I asked the Prisoner how she came to do it, and she said she was weary of her life, and wanted to be transported; and thought she could live better abroad than here, and gave me a slap on the face'.[39] The goods stolen were described as cotton and were the property of a woman of modestly comfortable means, able to afford a servant, living in a two-storey house, with a husband employed away from the house. All of these cases suggest a considerable breadth of consumption prior to the later eighteenth-century industrial expansion of the cotton industry.

Court records have a randomness that precludes quantitative analyses. It is impossible to state categorically what percentages of this sort of clothing or that type of fabric were contained in wardrobes representative of various social groups. But the very randomness in itself is a priceless asset, for it is not exclusive to one group or social ranking: visitors from around the country, seamen, merchants, nobles, artisans, labourers, tradesmen, and professionals were all players in the court annals. Nowhere else is there preserved a better general record of what people owned, the goods they had bought, stolen, acquired, and fought-over. The Old Bailey and other provincial assizes courts were the stage on which was decided rightful ownership of possessions with a cast made up from all strata of society. Time spent observing the cases parading on and off that stage impart an appreciation for the unusual and a recognition of the commonplace. Reports of cotton clothing appeared in transcripts throughout the early and middle decades of the eighteenth century, but in fewer numbers than the most popular and common varieties of apparel, such as linen and certain sorts of light wool textiles.[40]

---

[37] *Old Bailey Records*, May 1732, 138; Dec. 1741, 14.
[38] Ibid., Apr. 1741, 11; July 1742, 2.   [39] Ibid., Sept. 1745, 215.
[40] See Appendix 2 for a listing of apparel owned by men and women from a variety of occupations.

An early inventory of a labouring-woman's clothing survives due to the acquisitiveness of the lodger in the next room. In 1745 Mary Owen stole 'a cotton gown, val. 10s. a muslin handkerchief, val. 4d. a muslin hood, val. 3d. a cap, val. 3d. a flannel petticoat, 12d. and a handkerchief 4d.'.[41] This could well have been a best outfit, saved for Sundays and holidays. A slightly later and somewhat different example of the common working-woman's dress appears in a case from 1758. Mary Lewis was occupied outside weeding the garden and found her clothing missing when she came back indoors. She lost a linen gown, a linen handkerchief, a shift of unspecified material, a check apron, and a silk hat.[42] Neither inventory differs greatly from the other except in the matter of the fabric composition of the clothing. Mary Stinson kept house in Marsh Yard, Wapping, in 1743 and the acquisitive Rachael Milford found easy access to her wash-house where the clothes were being soaked in a tub. A cotton gown, a dimity petticoat, and an apron vanished.[43] A more detailed description of a working-woman of the same period provides added particulars on the sort of dress that would be common in many of its features to that of other women. She was described as wearing an 'old patched twill petticoat, calico red and white bedgown, old red, short cloak, white holland gown rough dyed, black velvet hood, black silk bonnet'.[44] Even in this attire there is a suggestion of interest in appearance.

As the mid-century approached a regular portion of cotton clothing became more the norm, although wardrobes could still be found devoid of any popular cottons.[45] None the less, the attraction of cottons took firmer hold over the 1740s, as illustrated in the garments almost lost by coal-heaver John Brown while enjoying some brew with his friend. A woman tried to carry some clothing out of the room, saying she was taking them to be ironed. Included in the inventory were two cotton shirts, one linen shirt, two linen aprons, and two linen shirts with ruffles; as Brown testified, 'one of the cotton shirts belong to Joshua Fox [with whom he had been sharing the ale], the other to James Richardson, he lodges in my house'.[46] In December of 1742,

---

[41] *Old Bailey Records*, Oct. 1745, 237.
[42] Ibid., May 1758, 201.
[43] Ibid., Apr. 1743, 155.
[44] *Ipswich Journal*, 17, Mar. 1759, quoted in Anne Buck, 'Variations in English Women's Dress in the Eighteenth Century', *Folk Life*, 9 (1971), 18.
[45] *Old Bailey Records*, Dec. 1742, 10.
[46] Ibid., Oct. 1749, 157.

William Taylor, a butcher, dressed warmly and stylishly in what appear to have been his best clothes and went for a walk on a Sunday to drink with friends on the Tyburn Road. Returning at 3 o'clock the next morning to get ready for work on Monday he was held up and all his clothes stolen. Taylor lost 'one Fustian Frock with 12 Plate Buttons, Value 40s. one white Duffel Coat, Value 15s. one Cloth Waistcoat, Value 5s. one Pair of Bucksin Breeches, Value 10s. one Perriwig, Value 10s. a Hat, Value 5s. and a Pair of Silver Buckles, Value 6s.'.[47] In this instance, as in others, the products of the cotton industry were common additions to a daily attire that was still largely composed of wool, linen, leather, and silk materials. But it is also interesting to observe that many of these collections of clothing display an interest in fashion. The problem for the cotton industry was to provide more of the sort of clothing people wanted at a cheaper price, offering inexpensive fabrics and hosiery that enabled even those of low income either to add minor items of dress to their existing wardrobe, like aprons, handkerchiefs, or ruffles, or to change their apparel completely.

For the skilled labourer and trained craftsman or clerk the more inexpensive materials permitted the undreamt-of luxury of several changes of clothes, of a sort that could be restored to pristine perfection through judicious washing.[48] The advantage of cotton and linen lay in price, washability, and fashionability; but for a time the price-differential of the latter made it more appealing than cotton. Nevertheless, the evident attraction of one type of cotton cloth enticed Ann Cuthbert to steal nine yards of printed cotton from Mrs Bell's shop. The culprit was chased by Charles Calmer who suspected her of the theft and took rather drastic measures to retrieve the goods. 'By seeing a bulk under her petticoats, I said, I fancy you have got Mrs Bell's goods under your petticoats; so I forced her down upon the ground, and I could not find anything under her first petticoat: I felt under her second petticoat, and there I found this printed cotton . . .'.[49] Such was the assiduity with which some sought to acquire cotton

---

[47] Ibid., Dec. 1742, 25.

[48] The ease of washing characteristic of cotton and linen clothing was a boon to families and a spur to greater cleanliness. By contrast, heavily soiled and worn wool clothing had to be re-fulled, or re-dyed and fulled to restore anything like a respectable appearance. In London and the suburbs 'scourers' operated 'who clean Men's Cloaths etc.' (Campbell, 201). Clothes-dealers also used the services of a 'scourer' to revive a wool coat, which might also include the services of a dyer, another tradesman frequently appealed to to refurbish a worn article of apparel.

[49] *Old Bailey Records*, Feb. 1748, 88.

textiles. By 1765 even a person described as a 'poor Woman' was able to bemoan the loss of 'a small running sprigged Purple and White Cotton Gown, washed only once, tied down with red Tape at the Bosom, round plain Cuffs, and the Bottom bound round with broad Tape'.[50] But the choice and volume of cottons available for the paying customer did not yet meet all needs. As late as 1776 one merchant wrote that in his opinion it would 'take a considerable time before the cotton manufacturers will fully supply the demands of the home trade'.[51] In an effort to expand the variety of textiles, production was redoubled in the last quarter of the century, as can be seen in the volume of retained raw cotton (Table 2.4). At least twenty-five sorts of cotton checks were being produced regularly between 1775 and 1785, along with approximately fourteen linen checks. Only two of the cotton-linen checks that twenty years before had been the staple of the industry were now manufactured. Pure cotton fabrics predominated by the last quarter of the century. The price of the cotton checks fluctuated throughout this period: starting from 13¾ pence per yard for yard-wide checks in the 1750s, up to 14½ pence in 1764, with a fluctuating price in the 1770s. By the next decade the price of cotton checks had dropped down and stabilized, holding around 12 pence per yard. Moreover, as more entrepreneurs began to operate spinning mills the price of all cotton goods fell throughout the remaining years of the century.

Year by year new cottons were brought on to the market. All manner of corduroys—Brunswick, Queen, Prince of Wales, and Genoa, corded Tabby, Wild Boar Tabby, corded thickset, as well as ribdeleur, ribdurant, everlasting, and Royal Ribb—issued from the manufactories of Manchester and vicinity between 1775 and 1785. Along with the abundant stocks of new fustians came an equal assortment of plain-surfaced textiles. In addition to the very successful checks, plaid cottons enjoyed a huge popularity and were offered in many designs and colours in at least twenty-two different widths, qualities, and patterns. Cotton sheeting also appeared in quantity for the first time, along with cotton towelling, the former available in approximately sixteen varieties and the latter in one. In addition to the standardized products, the industry was also better able to fill special orders to its customers' specifications, such as the moree fabrics 'flowered in the

---

[50] *The Public Advertiser*, 19 Feb. 1765.
[51] BM, Liverpool MSS, cliii, 38, 342, quoted in M. M. Edwards, *The Growth of the British Cotton Trade, 1780–1815* (1967), 27.

loom'.[52] (For a list of textiles identified as having been introduced by 1785 see Appendix 1.) The 225 cotton fabrics listed in Appendix 1 are illustrated of the productive capacity of the industry. They are by no means the sum total of production. Recently a sample book of cotton textiles from the Manchester merchants John & Benjamin Bower came to light. The examples of fabrics included in this volume represented the stock available in the mid-1770s from another agent of the British cotton industry. One hundred and twenty-one sorts of cotton, cotton-linen, and linen checks appeared in this folio; checks both for clothing and furnishings. Along with the checks were seventy-eight sorts of holland stripes, eighty-four cherydery, and 192 kinds of fustian. The total number of textiles listed was 476, an impressive indication of the acceleration of production within the industry over the previous quarter century.[53] Moreover, the price breakdown of the various textiles listed in the appendix confirms the focus of the industry's market emphasis.

The 225 cotton textiles culled from several sources were divided into three price-groups; those under 12½ pence per yard; those from 13 pence to 2 shillings; and finally those cottons costing from 2 shillings to the top-priced fabric of 11 shillings and 6 pence. The assortment of cotton textiles is in itself an impressive commentary on the developing productivity: viewed with reference to the three categories, the basis of growth becomes even more apparent. Of the 225 sorts of cottons, 126 fall within the medium price-range of fabrics, while the lower price-range contains fifty-five sorts of cottons. Forty-four sorts are listed in the medley of high-priced goods, and these are comprised in the main of hard-wearing, corduroys, velverets, and the like. The vast majority of the cotton, manufactured fall within a common category of affordable merchandise with the essential qualities of attractiveness, relatively low cost, and washability. Even when the technological progress enabled the manufacture of fine muslins and calicoes these were not produced exclusively in the highest qualities, but in a variety of qualities both medium and fine.

S. D. Chapman's assessment of the consumption of printed cottons found the same pattern of production and consumption. Nearly 50 per cent of printed cottons consumed in 1797 sold for between 12½ and

---

[52] Eng. MSS 1192, John Rylands Library, Manchester; *The Gazetteer and New Daily Advertiser*, 2 Aug. 1777; *The Beekman Mercantile Papers*, iii, 1, 115–17.

[53] Mary Elizabeth Burbidge, 'The Bower Sample Book', *Textile History*, 14: 2 (1983), 213–15.

18 pence per yard. This amounted to 12.6 million yards of printed cloth. The patterned fabrics in the second most popular range fell between 18¼ and 2 shillings and 6 pence per yard, and of these 5.4 million yards were sold, 21 per cent of the total. The cheapest printed goods, under 12 pence per yard, sold 2.8 million yards and accounted for 11 per cent of British consumption. In all, nearly 60 per cent of printed textiles consumed in the home market cost less than 18 pence per yard.[54] Cotton textiles of all weights, qualities, and varieties appeared in unequalled numbers from 1760 to 1800. What is now required is an assessment of the standards of dress of the various ranks of people in an effort to see how the components of dress and the standards of clothing changed over the period, from ordinary seamen to the hopeful daughters of the middle classes, from the hospital nursing assistant to the noblewoman of ample means.

CONSUMPTION OF COTTON CLOTHING, AND THE WORKING POPULATION

Contemporary comments, contents of shop invoices, and further itemized accounts of clothing present a graphic and diverse portrait of the costume of a cross-section of Britain's working people. Domestic servants maintained the most fashionable appearance of all the wage-earners, a factor long noted by contemporaries and historians. (See Plate 13, 'The Mop Trundler', for a contemporary illustration of the stylish servant. Note the white stockings—cotton?) As those in closest proximity to the wealthier middle and upper classes, the fashions currently in vogue were always in evidence. Emulation was standard practice, made easier after mid-century by the abundance of checks, printed cottons, and a growing assortment of other goods. By 1775, as we have seen, the choice in cheap, colourful cotton had grown dramatically. Records of dress and contemporary comment both indicate the importance of cotton textiles in the costume of female servants. More than one writer agreed with the sentiments of the commentator in the *London Magazine*, decrying the 'luxury of dress of our female servants, and the daughters of farmers, and many others, in inferior stations, who think a well chosen cotton gown shall entitle them to the appellation of young ladies'.[55]

---

[54] S. D. Chapman and S. Chassagne, *European Textile Printers in the Eighteenth Century* (1981), 90. [55] *The London Magazine*, 52 (1783), 129.

Inexpensive, brightly printed cotton textiles attracted ever-more consumers, as these textiles would contribute considerably to personal allure. The items stolen from a female servant in 1777 illustrate this vibrant sort of cotton clothing. Stolen were 'one pink and black Manchester gown, lined with green stuff; one garnett stuff gown; one striped and flowered cotton; one flowered cotton gon [sic]; one coloured quilted stuff coat'.[56] Not all of this servant-girl's clothes were made of cotton fabrics, but three out of four gowns originated in the cotton industry. The six or seven shillings needed to purchase sufficient cotton cloth to make a gown, or the eight shillings for a ready-made gown were low enough to create a potentially vast market among working-women, for whom these prices meant perhaps one week's wages or less. The list of apparel stolen from two daughters and one female servant of a brandy merchant living in Westminster illustrates how closely the wardrobe of the two classes of women resembled each other in content, if not in volume. Margy Read lost a check apron, a piece of cotton, a satin petticoat, and a muslin cap. Her sister Mary found a silk petticoat, a cotton gown, a pair of stays, a calico apron, and a linen petticoat missing from her wardrobe, while the servant, Mary Dentry, missed a silk cloak and a cotton gown.[57]

The belongings lost by servant Elizabeth Croudy in 1782, in the course of her move to a new position in Chancery Lane, echo the attention to dress characteristic of servants in almost every article. Included among the contents of the box which was stolen were 'seven yards of cotton for a gown ... two cotton gowns ... three dimity petticoats ... a muslin apron edged with lace ... a muslin handkerchief edged with lace ... a worked muslin handkerchief ... six muslin handkerchiefs ... and three pair of cotton stockings'.[58] Elizabeth Croudy had accumulated an impressive wardrobe, of which the cotton items comprised a part. (See Appendix 2 for a full listing of clothing owned over this period.) Mary Taylor was employed as a 'servant to a milk-woman', and also lost her box of clothing which contained:

one dark coloured cotton gown, value 14s. one cloth cloak, value 8s. four caps, value 4s. two pair of stockings, value 2s. three pair of worsted stockings, value 1s. one black silk handkerchief, value 1s. one muslin handkerchief, value 1s. one linen handkerchief, value 6d. one pair of shoes, value 6d.[59]

---

[56] *The Gazetteer and New Daily Advertiser*, 8 Nov. 1777.
[57] *Old Bailey Records*, July 1784, 900.
[58] Ibid., Feb. 1781, 131.   [59] Ibid., Feb. 1785, 497.

The ubiquitous cotton gowns and stockings, check aprons, and muslin handkerchiefs marked the progress of consumerism among the working population. This level of consumption did not begin with the establishment of spinning mills, rather it built on an established pattern of consumption, some of the highest levels of which were first seen, among the working population, in the purchases of servants. Take, for example, the account of goods owned by the servants of Mr John Andrews, of Newman Street, off Oxford Road. The newspaper recounted the goods stolen from the coachman's room over the coach-house:

four Irish Shirts marked S. T. numbered 2, 4, 5, 6, two Neckcloths, one Marked T. M. one fine Callico Shirt, the Ruffles half off; one scarlet Cloth Cardinal, one black Silk figured ditto, two Cotton Gowns, one a white Ground, with running Sprigs and Flowers, the other a dark Purple, with white Strawberries in large Diamonds, one white Silk Hat lined with blue and blue Ribbons.[60]

These goods were owned by this couple in 1766 and reflect their preference for the best quality of clothing they could afford. They were not unique in their preference, just more attuned to existing fashions as a result of their occupation, with enough money to spend on the apparel they knew to be in fashion.

The ordinary, everyday working-people of Britain dressed themselves in cottons with greater and greater frequency—a qualitatively new phenomenon. Comfort Bosley, a spinster living in a single room in London, had in addition to the clothes she was wearing a dark cotton gown and shift, plus two other cotton gowns, two linen aprons, one muslin and one cotton handkerchief.[61] The wife of lighterman Joseph Barber, of Christ Church, Surrey owned two cotton gowns in 1777, 'one a hop pattern, the other purple and white', the popular colours that year. A thief was described as wearing 'a yellow spotted waistcoat of silk and cotton'; a woman wanted by the law was last seen dressed in 'a yellow striped Cotton Gown'; the missing young woman from the country sought by her friends, was wearing at the time of her disappearance 'a cotton gown, with a purple ground, red and white spots'; and 'two notorious shoplifters' who escaped from Newgate wore a black-and-white striped silk-and-cotton gown, and a dark cotton gown with sprigs, respectively.[62] A fleeing slave attempted to

[61] *Old Bailey Records*, Oct. 1763, 257.   [60] *The Public Advertiser*, 24 Jan. 1766.
[62] *The Gazetteer and New Daily Advertiser*, 16 Jan. 1778, 8 Dec. 1777, 11 Dec. 1777, 23 Mar. 1779; *The Public Advertiser*, 19 Aug. 1784.

find a haven in London, in spite of the description published by her owner:

> A Negro Girl, about fourteen Years old, well featured ... and lately come from Philadelphia, speaks very good English, took with her two Jackets, one a blue and white Check, the other a dark Cotton, and two Gowns, one dark and one light flowered Cotton, and she is private Property.[63]

A resident of Islington bemoaned the loss, meanwhile, of a 'blue, white and chocolate striped Manchester gown, trimmed with white blond [lace]'.[64] Women in almost every area of work aspired to the calico gown, or checked handkerchief, brightening their appearance with items like the 'flowered Cotton Gown, Strawberry spotted'.[65] Tobias Smollet noted the characteristic gaiety of the holiday dress of young countrywomen whom he described in 1762, 'in their best apparel, their white hose and clean short dimity petticoats their gawdy gowns of printed cotton; their top knots and stomachers bedizened with bunches of ribbons of various colours'.[66] (See Plate 14, the Strawberry seller. Her pattern of attire is suggestive of the attention paid to dress by many working-women.)

The wearing of cottons spread out widely across the population. If one can visualize changes over time, they would have included a more vibrant, kaleidoscopic flourish of colours and patterns across the spectrum of the population. The reds, purples, yellows, greens, blues, dots, diamonds, flowers, sprigs, spots, and checks startle the eye and invite the attention not only of contemporaries, but also of historians, to witness the visible transformation unfolding among the people of Britain. That there should have been such a passion for colour is not surprising, nor is it surprising that this popular fancy elicited comments from contemporaries. English women were known for their love of vivid scarlet cloaks. In the eighteenth century the world of the working classes was permeated by many more articles of colour, in a manner that had previously only been an option for the affluent.

Among the working-men of Britain greater numbers of fustians and light cottons appeared in descriptions of their clothing from the 1760s onwards. The lighterman previously mentioned, Joseph Barber, still

---

[63] *The Public Advertiser*, 15 Jan. 1763.
[64] *The Gazetteer and New Daily Advertiser*, 5 Dec. 1777.
[65] Ibid. 17 Jan. 1763.
[66] T. Smollet, *Sir Lancelot Greaves*, quoted in C. W. and P. Cunnington, *Handbook of English Costume in the Eighteenth Century* (1964, repr. 1972), 324.

wore buckskin breeches and so did his sons, as too did Thomas Partridge, a wheeler's apprentice about the same time.[67] But within a space of five years, velverets, corduroys, thicksets, and velvets became the most frequently seen fabrics in men's outer garments. In 1784 the stableman's breeches stolen by his workmate were made of corduroy. A carpenter, John Parker, used a thickset coat, while three watermen, when attired other than in their trade costume, wore a range of heavy cotton garments: Manchester waistcoat and brown, velveret, or black Manchester breeches.[68] A bachelor, most probably employed as a skilled labourer, reflected in the contents of his wardrobe the growing preponderance of cotton textiles in daily use by working-men. Joseph Coleman possessed in total:

four cloth coats, value 20s. one velvet waistcoat, value 3s. one nankeen waistcoat, value 3s. one linen waistcoat, value 2s. one pair of leather breeches, value 5s. two pair of fustian breeches, value 3s. one pair of nankeen breeches, value 3s. one linen neck handkerchief, value 1s.[69]

Throughout the eighteenth century new sorts of woven and knitted cottons were added to the various categories of everyday clothing. A portrait of the average man at the beginning of the period compared to another at the end would reveal many alterations in the substance of dress, where cottons usurped the place of many traditional materials. Simplification in the style of men's clothing worked to the advantage of the cotton manufacturers. The early success of the cotton corduroys, velverets, and velveteens began the process, laying the groundwork for the progressive substitution of heavy, densely woven, napped cottons for both wool and leather clothing. Wool-stuffs, serges, worsteds, and broadcloths had been worn in coats, cloaks, jackets, and waistcoats for generations. None the less, by 1780 significant incursions had been made upon the market for everyday wool clothes, once so prominent a feature of the British costume. Viscount Torrington recalled in his diary the change in the demeanour of the tenantry from that of decades past, noting that 'in Bedfordshire an old tenant of my brother's, who wore the same coloured coarse cloth all the year round and tied his shoes with thongs', was succeeded by a son, who when the viscount

---

[67] *The Gazetteer and New Daily Advertiser*, 16 Jan. 1778; *Old Bailey Records*, Mar. 1774, 20–1.
[68] *Old Bailey Records*, July 1784, 1,114, Sept. 1785, 1,201; *The Public Advertiser*, 25 Oct. 1784.
[69] *Old Bailey Records*, Sept. 1784, 1,131.

came to call in 1776, received him in the parlour unexceptionably attired and offered his guest a glass of wine.[70]

The many sorts of napped, ribbed, and plain fustians invaded much of the market for men's waistcoats, coats, and frocks by the end of the 1760s, among labouring and middling as well as more affluent male consumers. Abraham Dent's ledger of credit sales listed various sorts of cotton clothing-fabrics sold from his shop to customers in his own and the surrounding villages of Cumbria, at about the time of this shift to heavy cottons. In 1762, for instance, the 'Rev. Mr Hodgeson at Brough' bought a fabric called 'Everlasting', a type of sturdy cotton corduroy, as did 'John Haygarth in town'. Dent himself chose a quantity of this fabric for his own use in 1764. Others in the village chose fustian, thickset, and a 'stript cotton'. The assortment of cotton fabrics sold by Dent was not extensive. However the half-dozen or so cottons he stocked suggest the effectiveness of this type of textiles in the home market, even in the less accessible areas of Britain.[71] About eleven varieties of fustians manufactured between 1775 and 1785 cost less than 12½ pence per yard and approximately thirty-four sorts of fustian were priced from 13 to 24 pence per yard. Corduroys, jeans, nankeens, erminetts, thicksets, corded tabby, and jeannette were some of the range of heavy cottons available from low to medium price. The cotton manufacturers also offered about thirty sorts of corduroy, velveteen, and velverets for from 2 shillings to 11½ shillings per yard.

Breeches made from the assorted cotton fabrics intruded upon the established market for leather and buckskin. As the third quarter of the century appeared, leather breeches were still commonly worn: two felons were sought in 1770, for example, wearing articles described variously as 'greasy Leather breeches' and 'Buckskin Breeches'.[72] However, greater production in the cotton industry soon allowed retailers to offer cotton breeches of many kinds and at low prices. The switch to cotton breeches was an innovation in men's clothing, painful to some established interests. The declining trade of one such leather-breeches maker was chronicled by his apprentice, Francis Place. During the term of his apprenticeship in the later 1780s, Place saw the

---

[70] G. B. Andrews (ed.), *The Torrington Diaries* (1936), i, 217.
[71] WBD/63/3a, Cumbria Record Office, Kendal. Note also the clothing of Thomas Fox, a Wolverhampton ironmonger, in 1745 and Abraham Baker in 1758, in Appendix 2.   [72] *Jackson's Oxford Journal*, 24 Mar. 1770, 31 Mar. 1770.

triumph of the new fabrics over the traditional sorts of leather apparel. Place's master was reduced to making cheap ready-made leather breeches for the Rag Fair, but as Place records, 'The Rag-fair breeches trade was rapidly declining in consequence of the increase of the cotton manufacture, corduroys and velvateens were now worn by working men instead of leather'. Several years later the prospects of this trade were even worse: 'the Leather Breeches trade had declined considerably and there was not nearly enough employment for the journeymen. Gentlemen rode in Corduroy and Cassimere Breeches, and leather was no longer commonly worn by any class of persons'.[73] Indeed the London proprietor of a clothing warehouse advertised, 'Breeches of various kinds, of the Manchester manufactory, such as ribdeleur, ribdurant, barragon, sattinet, everlasting, corderoy, jennet, stockinnet, etc. etc. all of the best kinds, at 16s. per pair'.[74] This was a commercial challenge whose momentum was impossible to curtail. Place's testimony confirms the profound alterations in the material composition of men's clothing.

Leather breeches appear in records only sporadically in the last decades of the century, while each new year beheld the arrival of additional varieties of corduroys, thicksets, and other fustians. Genoa corded velveret and Prince of Wales corduroy were some of the rugged cotton fabrics introduced in the late eighteenth century; while by the turn of the century Admiral Lord Nelson had his name attached to the new Nelson corduroy, symbolizing the strength and durability of the material. The inventory of goods sold to the Welsh smith John Powell typified the clothing of British working men of this time. These items were bought from the Morgan family who served the valleys and villages of South Wales with drapery and other products. From 1791 to 1801 the Morgan draper's ledger catalogued sales to Powell of corduroys including the new Nelson cord. Powell showed a clear preference for the hard-wearing, easily laundered cotton. Even the local tanner in a Welsh village evinced a liking for corduroy and velveteen for his clothing.[75]

At first working-men restricted their purchases of cotton clothing to those fabrics suitable for outer wear—waistcoats, frocks, and breeches— although check shirts too had been worn among the working classes

[73] Mary Thale (ed.), *The Autobiography of Francis Place* (1972), 80, 110.
[74] *The Gazetteer and New Daily Advertiser*, 2 Aug. 1777.
[75] D/D ma 139, Morgan Draper's Ledger, 26, 165, 334, Glamorgan Record Office, Cardiff.

for at least a century.[76] In the last decades of the century further alterations in labouring men's dress were noticeable. An example of this change appears in an account of a Liverpool sailor's clothes stolen in 1784. The seaman's attire consisted of 'one woollen jacket ... one man's hat ... one cotton handkerchief ... one striped cotton shirt ... one pair of silver buttons ... one pair of leather shoes'.[77] Since the sailor managed to hold on to his trousers even while in a drunken stupor these were not included in the inventory of his clothing. The significance of this list lies in the use of cotton for shirting. This was another notable innovation in the history of working costume, one preceded by the use of checked fabric in labourers' shirts earlier in the century. Mariners were probably the first working-men to use cotton shirts, examples of which appeared in records from 1732, when Benjamin Cook found his cotton shirts stolen from on board ship.[78] This sort of shirting was routinely used by those *en route* to the Caribbean or the Orient, where temperatures dictated lighter clothing. Cotton shirts were not as generally acceptable in Britain, although they could be bought ready-made in the 1770s. However, in the following ten years cotton clothing of this sort became more habitual for the working-man. Thomas Reed, a London hairdresser, appeared to have a decided preference for this sort of material as he numbered sixteen calico shirts among his possessions in 1784. London shops routinely carried men's cotton shirts, checked and plain, as their trade-cards and advertisements attest, and this item ceased to be a novelty. Indeed, William Dowell offered blue and white cotton shirts for sale from his shop in Deptford.[79]

Ready-made cotton shirts were also listed among the merchandise sold by Morgan the draper, and one can assume that they would be in general use throughout the country. Thomas William, a servant at Dufferin, spent 14 shillings and 2 pence on two shirts in 1791. Four years later Francis Yorrath bought a cotton shirt for only 5 shillings, a common choice with numerous characteristics to its advantage. As cotton production increased the price of cotton fell. Simultaneously articles of cotton clothing were included in the massive purchases

---

[76] Appendix 2 provides examples of these kinds of working-men's attire, for example, the cases of Christopher Bones, John Ward, William Grace, and John Manning, who were respectively a servant, watchman, lodging-house keeper, and locksmith.
[77] *Old Bailey Records*, Sept. 1784, 1,185.   [78] Ibid., May 1732, 138.
[79] Ibid., Dec. 1784, 28–9; Q/SB Examination and Information, 24 Oct. 1787, Kent Archives Office, Maidstone.

commissioned by the armed forces. Part of the enthusiasm for cotton clothing can be attributed to the decline in price of shirting fabrics; an equal measure of appeal can be attributed to the intrinsic characteristics of the fibre. The naval physician Thomas Trotter expound new ideas of health and hygiene for working-men when he denounced the practice of issuing flannel clothing to be worn next to the skin. Trotter declared that it turned His Majesty's seamen into 'walking stinkpots'. He insisted that, 'If British seamen are to wear flannel next the skin, they ... must soon lose the hardiness of constitution that fits them for duty. Clothe them as warm as you please but in the name of cleanliness give them linen or cotton next the skin'.[80] It cannot be deemed a coincidence that large orders for cotton shirts were placed by the navy in 1793, a trend in consumer preference for cotton clothing that was presaged by the choice of earlier seamen and mirrored the trend among the civilian population as well.[81]

In conjunction with the sartorial revisions in coats, frocks, waistcoats, shirts, and breeches, cotton hose had also found a growing market following its introduction years earlier. As with textiles, the assortments of cotton hose increased in quality and quantity as the craftsmen cultivated the home market and profited from the ongoing rage for light hose. By the 1770s over fifteen sorts of cotton hose were ordered from a British supplier, where twenty years before there had been many fewer alternatives.[82] In 1736 *Read's Weekly Journal* claimed that 'White stockings were universally worn'.[83] This assertion most probably referred to the ranks of the genteel and their disciples. Whereas Mrs Papendick's notation reflected the qualitative consumption of cotton stockings when she wrote in 1789 that, 'Black worsted stockings now only seen on servants of an inferior order and the lower working classes'.[84] The prosperity of yet another segment of the cotton industry came about through the popular demand for cotton hose, wherein tradesmen like Morgan sold to customers from as diverse circumstances as captain and currier, surgeon and ironmonger, tailor and tanner. The Welsh draper sold plain cotton hose, ranging in price from 17 pence to 5 shillings per pair; he also stocked grey and ribbed

---

[80] Thomas Trotter, *Medicina Nautica* (1801), iii, 93–4, quoted in Peter Mathias, 'The armed forces, medicine and public health', in his *The Transformation of England* (1979), 278.   [81] Adm. 49/35, Public Record Office.

[82] Invoices, Peach & Pierce, 10 Sept. 1770, Beekman Mercantile Papers, New York Historical Society.

[83] *Read's Weekly Journal*, 1736, quoted in Cunnington and Cunnington, *Handbook*, 174.   [84] *Mrs Papendick's Memoirs*, quoted in Ibid. 396.

cotton stockings.[85] From the valleys of South Wales, across the provinces, and through the metropolitan centres cotton stockings were staples.

The customers identified by trade or occupation in the Morgan draper's ledger bought cotton textiles and clothing as regularly as did their counterparts in England. Examples of some of the regular consumers are the maltster David Hopkins who, as well as buying printed calico and love ribbons, bought himself a scarlet swansdown waistcoat shape and a length of wild boar tammy, both products of the Lancashire cotton industry, despite the exotic-sounding names. Employers purchased ready-made cotton gowns for their servants. Mr Rees Howell, the tanner, bought muslin and a quilted petticoat for his wife. Other skilled craftsmen, like Evan Evans the cordwainer and William Storey the joiner, spent between £10 and £20 annually on textiles and textile goods from Morgan. A substantial portion of these expenditures was on cotton textiles for clothing of one type or another. The tailor in Neath, Richard Daniels, also bought several waistcoat shapes, cotton hose, and printed calico, as well as velvet, velvet collars, fustian, plain cotton, and check, over the four years from 1796 to 1800. The baker, the millwright, the tanner Philip Buttrel, and the currier, Mrs Rachael Morgan, all purchased the same sorts of merchandise, although the many varieties of cotton textiles allowed individual tastes free reign within a price category.[86] Even in the humble occupation of parish watchman the incumbent dressed himself in nankeen breeches to face the damp and fogs of the London night.[87] Skilled and unskilled, tradesman and craftsman, were accumulating and replacing clothing at a steady rate, and the proliferation of cotton textiles contributed to this increase in consumption.

There seemed to be no limit to the diversity of goods that could be manufactured from cotton yarn. Fabrics were made providing consumers with a host of contradictory features from the same basic material, depending on the style of yarn spun, the structure of the weave, and the method of finishing. Cottons were made that were warm or cool, heavy or light, hard-wearing or ephemeral, smooth or napped. Manufacturers created one product after another, as fast as technological problems could be overcome, most in imitation of existing merchandise but with the added feature of easy care

[85] D/D ma 139, 66, 92, 117, 134, 144, 162–3, 245, 255, 263, Glamorgan RO.
[86] Ibid. 1, 26, 29, 117, 165, 187, 248, 263, 303, 334.
[87] *Old Bailey Records*, May 1790, 531–2.

characteristic of cotton clothing. It is rare to discover an inventory from among the labouring classes which does not include some item of cotton clothing by 1780, and by the close of the eighteenth century cotton textiles and knitted goods were conventional commodities in the clothing of the working population. Appendix 2 offers further examples of the possessions of these men and women, that can be contrasted with the collections from the wealthier classes. Among this group there was a consistent and growing accumulation of clothing in the latter decades of the century, and in particular cotton clothing. There is no better indication of the omnipresent cottons than the list of clothing stolen from Mary and John White. This catalogue included articles owned by the couple, as well as the neighbours' garments brought for washing. The neighbourhood of row houses in Bromley described by the victim would not be likely to generate unique or unusual habits of consumption. Rather, the commodities such as calico bedgown, muslin aprons, calico aprons, muslin tucker, and the like, reflect the commonplace.[88] The commonplace in this era was the greater use of cotton clothing. Broad inferences about the rate of consumption can be made using S. D. Chapman's calculations of the rise in the levels of consumption of printed muslins and calicoes in Britain from 1790 to 1800. During the last ten years of the century the number of yards of printed cotton sold in Britain almost doubled: 15.01 million yards of British printed cottons were sold in the country in 1790–1; ten years later nearly half the 28.69 million yards produced were consumed domestically. When the price-structure of the printed cottons is recalled—80 per cent of printed textiles in 1797 cost less than 2 shillings and 6 pence per yard; 59 per cent under 18 pence per yard—the extent of the dependence of this industry on the consuming habits of the domestic market becomes apparent. The largest portion of the cotton materials produced in Britain was the stuff of workaday clothing and Sunday best. The labouring people, artisans, tradesmen, and all those who aspired to gentility 'carried off' the bulk of British cottons.

Did the poor follow a similar path of consumption? In fact, although observers noted the visible difference in some types of clothing among the labouring classes—stockings in particular—even among the poor there was a more extensive use of cottons. In 1789 an anonymous lady published a lengthy plan of her own devising aimed at providing the

[88] Ibid., Jan. 1793, 305.

working poor with the essentials of dress at the lowest possible cost. Parents, two children, and a baby were included in this exercise whereby clothes would be made up by the Sunday school to be sold to the poor of the parish at cost. As the fabrics would be bought wholesale, this was thought to offer a considerable saving.[89] Women and children were issued checked fabric for aprons; the cost and description suggests this would have been cotton check. Approximately two yards were required for the largest girl's apron and double that for the largest woman's. Boys were equipped with sturdy white napped jackets, as well as 'red napped waistcoats'. The description suggests a fustian of some sort was deemed suitable for boys' rough and tumble. The father's costume was made without any cotton, although tapes and pocket materials would almost certainly have been used. The family member allotted the greatest amount of cotton was the baby. Included in the list of items thought essential were two sheets of Lancashire cloth for the wife's childbed; two baby shirts made of figured diaper, and two baby frocks made of printed cotton. The latter item was proposed to be made from printed cotton costing 2 shillings per yard, the sheeting to cost 14 pence per yard and the figured diaper 7½ pence per yard. The price per yard of these fabrics matched the prices for similar cloth listed in the ledger of cotton manufacturers and wholesalers. Baby would most probably wear cotton, according to this plan.[90]

The proposal detailed by the charitable Hertfordshire lady was intended to represent the most fundamental level of consumption of textiles judged necessary. Seen as the minimum standard of dress, the plan reflects changing patterns of expectation, both the expectation of the individual and those of society about the level at which it allows its dependents to live. The account of the wearing-apparel received and distributed by the constable of Cheshunt, Hertfordshire, from 1782 to 1786, provides additional indications of the basic standards of clothing for which the local authorities were willing to pay. Pieces of cotton check and dozens of check handkerchiefs were bought or donated to the parish authorities along with 'Yorkshire plain', linsey-woolsey, dowlass, flannel, and hessen, as well as used clothing. The supervisors of the poor do not seem to have been over-generous, but the

---

[89] BM 1609/1131, *Instructions for Cutting Out Apparel for the Poor* . . . (1789), iii–iv. I am indebted to Mrs Madeleine Ginsburg, formerly of the Textile Department, Victoria & Albert Museum, for drawing my attention to this source.

[90] Ibid. 40, 45, 56, 68, 72, 79–85.

necessities were more than evident.[91] The worthy poor were never provided with more than the perceived basics in clothing, so as to discourage sloth and a reliance on the parish. However, the minimum so described varied as the prosperity of the society increased regionally and nationally. Basic clothing, as conceived by societies' arbiters, appears to have been redefined. As levels of consumption rose alterations also occurred in the concepts of an acceptable minimum of clothing and household textiles. It can be inferred that those not receiving charity owned more in the way of clothing, for it was not seen as an extravagance to clothe the offspring of a family receiving parish relief in printed cotton costing 2 shillings per yard. That assumption on the part of the author of the plan is a pertinent commentary on the scale of societal expectation and the rising rate of consumption through all ranks of society.

'DRESS OF THE TIMES': MIDDLE AND UPPER-CLASS CONSUMERS

The consuming habits of the middling sectors of society are more easily deduced. This group was not reserved either in their selection of goods, or in their self-congratulation when at last the goods were bought. Being generally more literate, family records, accounts, sample-books, letters, and a miscellany of other paraphernalia allow a closer look into closets and clothes-presses over this period. Not surprisingly, the consumption of British cottons by this group differed from that of the labouring classes. A greater range of social activities necessitated that the tradesman, attorney, clergyman, officer, and their families own a greater number of costumes in approximately the current mode. The wives and daughters of the middle classes were not restricted to checks or cheap printed cottons for their daily attire, but had a wide choice of fabrics for the different sorts of garments that comprised their wardrobe (see Appendix 1). The versatility and variety of cotton textiles led to purchases of equally diverse sorts of clothing, such as 'a white calico [morning dress] trimmed with muslin, a second dark cotton trimmed with silk'; 'a blue, white, and chocolate striped Manchester gown trimmed with blond'; or 'a black and red striped Manchester stuff petticoat, trimmed for a negligee'.[92] The more costly fabrics attracted those more able to afford the price, especially if

---

[91] D/P29 9/1, Cheshunt constable account, 1782–6, Hertfordshire Record Office, Hertford.
[92] *The Gazetteer and New Daily Advertiser*, 3 Nov. 1777, 26 Jan. 1779.

original in their pattern or colour. Novelty was the most desirable quality in printed goods, and novelty was provided in plenty by manufacturers who well understood the criterion of their customers. Benjamin Franklin typifies the middle-class consumer intrigued by innovative fabrics. He sent a package of English cottons to his wife in 1758, describing the contents in the accompanying letter as '56 Yards of Cotton, printed curiously from Copper Plates, a new Invention, to make Bed and Window Curtains . . . Also 7 yards of printed Cotton, blue Ground, to make you a Gown'.[93] The middle classes need not have bedecked themselves in cotton every day or for every occasion, but the many new and attractive textiles devised over this period were to be an invaluable addition to fashionable wardrobe.

The Revd James Woodforde commented frequently on clothing that caught his eye and purchases of particular note were recorded in his diary. Woodforde generously provided for his niece Nancy, who made her home with him in Weston, Norfolk, and in the winter of 1780 he wrote that, 'Nancy had her new Cotton Gown brought home this evening from Norwich by Mr Cary and I think very handsome, trimmed with green Ribband'. Woodforde added, with a gentle smugness, that this was 'a Cotton of my Choice'.[94] The clothing of the middle classes was an advertisement and assertion of their status as much as a practical reflection of personal income. Society set certain standards of dress to which the aspirants to gentility had to adhere. These precepts sparked a greater consumption of the higher-priced cottons than necessity demanded. Growing numbers aspired to the standards of gentility; in the pursuit of gentility appearance was paramount. That being the case, expenditures were generated by the countless scores subscribing to the philosophy that, 'on our dress depends the general estimation of the world, as persons are every way looked upon according to their cloathes, and their merit valued by the judgement of their taylor or their mantua-maker, dress is a thing which deserves our serious consideration . . .'.[95]

---

[93] Benjamin Franklin, letter, quoted in Montgomery, 29.
[94] James Woodforde, *The Diary of a Country Parson, 1758–1802*, ed. John Beresford (1924), i, 299.
[95] *The Lady's Magazine*, 5 (1774), 538. It is interesting to reflect how qualitatively public perception of morality had changed. In 1714, Mandeville expressed his satirical views on the matter of dress, private greed, and public good, in terms that were very similar to those published in the above magazine; and at that time he was considered universally odious. At the end of the century the same comments are taken as a guide to careful consumption.

The composition of women's attire changed more drastically than that of the male among the middle classes, following the dictates of fashion. The indefatigable diarist Mrs Phillip Lybbe Powys noted such changes in the dress of the royal princesses at Ascot in 1795. The royal maidens were 'dressed' ... in the dress of the times, all clear muslin'.[96] When costumes made of cotton were donned by royalty, similar gowns would be worn in their thousands by fashion's disciples around the country.

The household accounts of one middling Lancashire family exemplify how families of this type translated their greater prosperity into mounting consumerism. The family's assumption of the trappings of the middle class are unmistakable in the expenditures for dancing-lessons, the conversation club attended by the wife, and the purchase of books such as the *History of England* and the *Dictionary of Trade*.[97] The years covered by the account-book are providential. Running from 1770 to 1782, they include the time of the greatest expansion to date in the cotton industry and coincided with this family's growing wealth. The first year records one purchase of cotton goods—10 shillings and 6 pence for fustian. None was noted in 1771 or 1772, while in the next year over £3 was spent on large quantities of furniture-check. The pattern of steady spending on household improvements continued until the next decade, at which time there began a flurry of shopping for the eldest daughter. Through the years 1780–2 the entries reflect a complete pre-occupation with fashionable clothing for the whole family, but especially for the daughter, Hannah. (See Plates 15 and 16, for an example of a child's embroidered muslin dress of the late eighteenth century.) Several pounds-worth of muslin were purchased in 1781, in addition to other fashionable items. In subsequent years as much as £5 were spent each year to augment the daughter's wardrobe, concentrating first on muslins for accessories and then on printed cottons for gowns.[98] The pattern of expenditure encapsulated in this family account-book mirrors a level of consumption that would have been common for the new generations growing up during this era.

The Barbara Johnson Sample Book recounts a similar middle class

---

[96] Emily J. Climenson (ed.), *Passages from the Diaries of Mrs Philip Lybbe Powys 1756 to 1808* (1899), 282–3.
[97] Eng. MS 989 R9 1695, Hibbert-Ware Papers, John Rylands Library, Manchester.
[98] Ibid.

pre-occupation with the selection of fashionable clothing.[99] Barbara Johnson kept a rare and detailed personal record of clothing-purchases over the last half of the eighteenth century and through to the third decade of the nineteenth. Barbara Johnson was born the only daughter of an English clergyman in the parish of Olney, Buckinghamshire. For a large portion of her life she lived in Northamptonshire on a modest income.[100] The comparative remoteness from the centres of fashionable society was no obstacle to Miss Johnson's study of the latest modes, and from about mid-century, at the age of 8, until her death in 1825, at over 80 years of age, she kept a scrap-book and clothing diary. Catalogued within its pages are lists of clothing purchased, along with squares of fabric pinned to the pages, often with the price per yard of the goods included. (See Plate 17, a page from the Sample Book.) The Sample Book compels close attention, first because of the pertinent decades covered by this volume and second because of the representative nature of both Miss Johnson and the clothing she purchased. Barbara Johnson was part of the middle ranks of British society. Family and friends brought her news of the London world and an occasional first-hand taste of entertainments. One can assume that both her interests and her patterns of consumption reflected common judgements and practice, as did her collection of printed illustrations that accompanied the fabric swatches. Significantly, over the eighteenth century, this volume illustrates the changing place of cotton fabrics in her wardrobe, the shift in the constituents of dress away from silk, wool, and linen towards cotton materials.

During the eighteenth century the majority of the dress-pieces bought for clothing were silk; thirty-seven in total. But of this total thirty-one of the lengths of silk were acquired before 1779; only six new gowns were made of silk in the last twenty years of the century. The cost of these silks ran from 12 shillings to 4 shillings per yard, though the bulk of her purchases of silk lutestring, tobines, and taffetas cost between 4 to 7 shillings per yard. Wool-clothing purchases were similarly concentrated in the earlier part of the volume. Nine wool garments were listed before 1763 and only one in the years running up to the nineteenth century, and that last was a gift to Miss Johnson. Another interesting feature of this record was the number of linen materials bought. In her journal the clergyman's daughter registered

---

[99] The Barbara Johnson Sample Book has recently been published. See Natalie Rothstein (ed.), *Barbara Johnson's Album of Fashions and Fabrics* (1987).
[100] Ibid. 10–14.

fewer lengths of linen for clothing than any of the other fabrics.[101] There were several sorts of linens bought in the early 1750s, one of which was inaccurately described as a printed cotton. A small flurry of printed linens appear just prior to 1780, after which there were apparently no further additions of linens to her collection of clothing.[102]

One of the earliest listed purchases of a British cotton by Barbara Johnson was fustian for a riding dress; this was bought in 1757. The choice of cotton fustian for this garment illustrates the initial penetration of the domestic market for both men's and women's clothes by heavyweight cotton textiles. Such fabrics provided a less-costly alternative to comparable wool textiles. The difference in price was substantial: Barbara Johnson paid 5 shillings and 6 pence per yard for the cotton twill cloth for her riding dress, while a wool broadcloth riding dress was nearly four times as dear, costing 20 shillings per yard. In cost and easy care the advantage went to the cotton fustian.[103] Cotton fabrics appeared in the sample book in the late 1740s and not again until the 1760s; from that time until 1779 they comprised a small but constant part of Barbara Johnson's new apparel. Printed linens served the same function as did the cottons, and for a while, before 1780, she appeared to favour the two sorts of textiles equally. Four lengths of printed linen were bought early in the 1750s for 2 shillings and 4 pence per yard, 2 pence less per yard than the printed cotton bought in the next decade. However, the next piece of linen noted in 1764 cost 3 shillings per yard and the three remaining sorts bought in 1779 continued to show a higher price. The last linens bought cost from 3 shillings and 3 shillings and 2 pence per yard.[104]

There was an obvious correlation between the improvement of light-cotton dress materials and the disappearance of many comparable linen, silk, mixed linen-wool and silk-wool fabrics from the pages of the sample book. The first notations of printed cottons in the volume were followed later in the 1760s by another sturdy cotton twill and next by a glazed cotton print; while over the next decade an embroidered Indian cotton and purple-and-white printed cotton were added to her complement of clothing—the latter being a particularly fashionable

---

[101] There is no mention of underclothes in this album. Linens would have been used in the making of these garments; thus one cannot assume that all Barbara Johnson's apparel was noted in this book.
[102] T 219–1973, The Barbara Johnson Sample Book, Textile Department, Victoria & Albert Museum, London, 2–21.   [103] Ibid. 4, 8.   [104] Ibid. 1–46.

colour-combination in the 1770s. Linen clothing-textiles enjoyed an advantage only as long as it took to develop cottons of the same quality or, indeed, cottons that offered added features such as the glazed cotton finish. Between 1780 and 1800, Barbara Johnson routinely purchased British-made cottons. The cheapest cotton cost as little as 20 pence per yard, though the average price of cottons ran from 3 to 5 shillings per yard. Nevertheless, these cottons were cheaper than the equivalent 4 to 7 shillings per yard for most of the silk fabrics bought. The price differential favoured the replacement of silk gowns with clothing made of cotton: witness the seventeen cotton gowns made between 1780 and 1800, compared to only six silk and one wool. In the last two decades of the eighteenth century some of the fabrics Barbara Johnson routinely noted included 'blue muslin', gingham, cotton tabby, calico, and chintz. She subscribed to the current modes. Thus her selection of cottons reflected equally common standards among the middling ranks.

Evidence from among the more affluent middle ranks confirms the appeal of cottons. Mrs Miers of Neath, a substantial customer for the Morgans, bought herself two 'Super Chintz Gowns' for 41 shillings and 6 pence in 1790. Over the next eighteen months Mrs Miers also made frequent purchases of cotton hose, both grey and white, plus six cotton caps. Mrs Jane Biby, a widow of comfortable means, was even more open-handed in her selections. At least once and sometimes twice each spring, from 1789 to 1802, she spent from 17 shilling to 1 pound and 10 shillings on various sorts of printed cottons, costing from 2 shillings and 4 pence to 3 shillings and 9 pence per yard. These expenditures were in addition to the cotton hose, cotton check, furniture-check, and velverets bought over the same period. The clothing inventories of Catherine Walker, gentlewoman, Sarah Bearcroft, widow, publican Thomas Gibson and his sister and mother-in-law, and the wife of Thomas Cruden, doctor of physic, reveal comparable clothing across the spectrum of middling occupations and positions.[105]

The men of the middle and upper strata exhibited an almost equivalent level of consumption of cotton clothing as their female counterparts. The *Old Bailey Records* note several kinds of cotton garments among the kit of naval officers passing through London. The first mate of the *General Elliot* lost three cotton waistcoats and ten pairs

---

[105] Appendix 2, nos. 22, 34, 35, 40.

of cotton stockings, in addition to his silk waistcoat and hose and two cloth coats. The captain of the ship *Venus* boasted a much larger collection of clothing. The items he lost included four pair of nankeen breeches, eighteen pair of cotton stockings, and eighteen pair of cotton trousers.[106] Even in the valleys of South Wales, professional men and gentry bought cotton clothing and textiles, goods that would have been commonly available throughout Britain, such as Wigan striped cotton, muslin, jean, genoa velvet, muslin handkerchiefs, and fustians.[107] The documentation of this period reveals a transformation of dress, a qualitative remodelling of expectations, attitudes, and ambitions. An observer of the evolution of dress of the period wondered at in the erasure of so many of the visible, distinguishing social features of dress and at the more democratic patterns of consumption in clothing.

Both in the make and materials of their garments is a union of simplicity and attention to convenience, and the difference between a man of first-rate elegance in an undress and a gay disciple of George Fox is often scarcely discernible. . . . To what rank has the rage for muslin not extended, both upward and downward?[108]

[106] *Old Bailey Records*, Oct. 1784, 1,281–2; 1,334–5.
[107] D/D ma 139, 66–7, 140, 166, 186, 243, 263, Glamorgan RO.
[108] *The Universal Magazine* (1810), quoted in C. W. and P. Cunnington *Handbook of English Costume in the Nineteenth Century* (1959), 13–14.

4
DISTRIBUTION AND SALE OF BRITISH COTTONS IN THE HOME MARKET

THE NATIONAL MARKET: TRADE AND TRAVEL

THE national market for British-made cottons in the eighteenth century was a barometer of change both broadly focused and narrowly specific. The distribution and trade in cotton textiles revealed the evolving characteristics of the national market for a popular commodity. The rate at which the market was served, the period by which a common pattern of demand was established, and the manner in which the domestic network functioned are all addressed in this study. The vigour of Manchester-based wholesalers and retailers limited London's role in the distributive network, consolidating the Lancashire centre not only as the hub of manufacture but also of distribution, an aspect not previously identified.[1]

Manchester assumed and sustained a pivotal role in the sale and distribution of cotton products. The efficiency with which markets were supplied led to the consistent availability of British cotton textiles throughout the country, a uniformity established around the 1760s and reinforced in the succeeding decades of industrial growth.

The marketing of textiles was never a sedentary trade, and this continued to be the case for tradesmen in the first half of the eighteenth century. Regardless of the hazards and inconveniences of travel, merchants and humbler traders covered miles of territory, over rutted tracks, through glue-like mud and the hazards of flooding streams and wheel-fracturing roads, collecting the requisite supplies of fabrics and bringing bales of cloth for sale at markets and fairs. Most

[1] A. P. Wadsworth and Julia de Lacy Mann considered that 'probably the greater part of the foreign and home trade in Manchester goods was carried on through London, then the focus of the distributive and overseas trade'. However, even this tentative assertion of London's pre-eminence seems too sweeping judged on the evidence then presented in their brief segment on the inland trade. The greatest number of their examples described the numerous Manchester Men and their significant contribution to domestic trade. *The Cotton Trade and Industrial Lancashire* (1931), 236–7.

were inured or at least resigned to the exigencies of their trade. Some travellers even discovered compensations for the bitter cold and precipitous roads, as did Daniel Defoe outside Rochdale on the road into Yorkshire. Writing in the middle of August Defoe noted with dismay:

the mountains covered with snow, and ... the cold very acute and piercing; but even here we found ... the people had an extraordinary way of mixing the warm and the cold very happily together; for the store of good ale which flows plentifully in the most mountainous part of this country, seems abundantly to make up for all the inclemencies of the season, or difficulties of travelling.[2]

Detours and delays were accepted as the middleman in the textile trade proceeded deliberately with their affairs. The pace would seem leisurely by later standards, but this rate of progress was not by choice but of necessity, for the collection, sale, and distribution of textiles depended on the weather, the state of the roads, the waterways, and the coastal seas. Defoe described the numbers of people employed in the business of this trade: 'multitudes of people employ'd, cattle maintain'd, with waggons and carts for the service on shore, barges and boats for carriage in the rivers, and ships and banks for carrying by sea, and all for the circulating these manufactures ... for the consumption of them among the people'.[3]

The business of Samuel Finney and his associate Ralph Davenport used many of the transportation methods enumerated by Defoe: coastal vessels, pack-horses, and wagons were all enlisted to extend their trade and serve their customers. Samuel Finney was a manufacturer and dealer in checks, fustians, tapes, and other assorted goods made in Manchester and neighbouring counties. He also solicited supplies of hose from London wholesalers that he then sent to Bristol. The market he served was the West Country and Welsh border region, using Bristol as a primary warehousing centre from which to send on supplies. Truro was the main base of operations in the Cornish peninsula.[4] Invoices and letters remain for the decade of the 1730s as evidence of an extensive business network, chronicling sales, distribution, and payment-practices that linked London merchants, Bristol, and Truro factors, in addition to numerous customers, many of whom were themselves in trade.[5] The distribution

---

[2] Daniel Defoe, *A Tour through the Whole Island of Great Britain* (Everyman edn., 1966), ii, 189.   [3] Id., *The Complete English Tradesman* (1726), 255.
[4] DFF/21, Letters and Invoices, 26 June 1732, 7 July 1732, 26 Oct. 1732, Cheshire Record Office, Chester. .   [5] DFF/21; DFF/20/1–64, Cheshire RO.

system devised in this period relied on coastal shipping in combination with land travel from the warehouse site to shops and trading facilities throughout the west of Britain. William Cox acted as the Bristol factor for Finney, forwarding orders to Truro and occasionally to Falmouth, as well as arranging other aspects of Finney's affairs, most probably for commission. The sea-borne character of Finney's commerce extended to periodic trading ventures to Holland, or to speculative purchases of coal along the Welsh coast. For the most part, however, coastal traffic was the means of moving Finney's large consignments of assorted Manchester materials from Lancashire to Bristol and Truro for later sales in the more distant ports, towns, and villages.[6] Defoe himself noted the vigour of the Lancashire trade in the mid-1720s and the ease with which business was conducted between the north-west and the West Country.

all this part of the country is so considerable for its trade, that the Post-Master General had thought fit to establish a cross-post thro' all the western part of England ... to maintain the correspondence of merchants and men of business, of which all this side of the island is so full. ... This cross-post begins at Plymouth, in the south west part of England, and, leaving the great western post road of Excester behind, comes away north to Taunton, Bridgewater and Bristol; from thence goes on thro' all the great cities and towns up the Severn ... thence by West-Chester to Liverpool and Warrington, from whence it turns away east, and passes to Manchester, Bury, Rochdale ... and ends at Hull. ... The shopkeepers and manufacturers can correspond with their dealers at Manchester, Liverpool and Bristol ... without the tedious interruption of sending their letters about by London ...[7]

Records from the 1730s describe an alternating agenda, wherein the manufacturers Samuel Finney and Ralph Davenport took turns travelling from point to point, to the villages and towns scattered along the valleys and margins of the coast. Each man apparently kept separate accounts, as well as participating in joint ventures, acting in the other's interest, providing supplies for each other, and collecting debts. In November 1731, Ralph Davenport dispatched a letter to Finney enquiring as to the volume of sales since his departure from Manchester. Included in this missive was a list of those owing money to Davenport, plus an exhortation to Finney to 'be as speedy in your geting [sic] in the outstanding debts as possible, for you know the pressing occasions I have for money att this time of the yeare'.[8] This

[6] DFF/20/22, Cheshire RO.   [7] Defoe, *A Tour*, ii, 188–9.
[8] DFF/20/22, Cheshire RO.

document was one of several that listed customers supplied by Finney and Davenport. The majority lived in Devon, Dorset, and Cornwall; however, this was not the entirety of their trade. Business-ties existed with such well-known families in the early cotton industry as Touchett and Philips,[9] as well as more than a dozen other merchants, some of whom also had names familiar from later merchant records, such as Livesey and Shaw.[10] But the majority of their trade was in the western counties of England; their joint prosperity depended on the skill with which they served this community.

Truro was ideally positioned to act as a secondary warehousing site, being slightly more than half-way down the length of Cornwall, at the head of the Truro River and the Carrick Roads, and at the point of intersection of the principal land routes on the peninsula. This town was described by Defoe as being 'a very considerable town' with 'a good port for small ships', as well as having a long established coastal tie with Bristol.[11] Once supplied from their storehouse, Finney or Davenport set out from Truro to the more westerly towns of Redruth, 'St Anns' [Agnes?], Camborne, Helston, St Keverne, St Ives, St Just, Penzance, Falmouth, St Mawes, Penryn, and back to Truro. The villages and towns to the east were also served from this point: Probus, Tregony, Mevagissey, St Austle, St Columb, Lostwithiel, Fowey,

---

[9] DFF/20/40, 42, Cheshire RO. The reference to 'N. Philips' provides evidence of the precursor to the later J. & N. Philips & Company. A. P. Wadsworth and Julia de Lacy Mann explained that the uncle of the founders of the company first had a Manchester-based business in the early eighteenth century, 'probably in smallwares'. The records of Samuel Finney lend support to that supposition and provide actual evidence of the business affairs of N. Philips, although in this instance they were not propitious. Finney was informed by Cox that a bill on Philips for £145. 14s. had been refused and returned by another merchant. The return of this bill caused Finney some pecuniary discomfort and professional chagrin, but this evidence provides needed confirmation of the antecedent of the J. & N. Philips & Company.

[10] A sort of business genealogy can become a recurring preoccupation when dealing with tradesmen's records such as these. Seemingly unconnected business accounts throw out references to names familiar from other documentation and very shortly one begins to detect a lattice-work pattern of commercial associations, tying manufacturers or tradesmen already identified into context with those more recently uncovered. This is a fascinating occupation in itself. But it also has a utility in the overall study of the cotton trade; for through cross-reference and identification of trading partners one may more accurately assess the status and the level of trade of a particular entrepreneur. An anonymous tradesman may thereby be placed in perspective with regard to the constituents of his trade as well as his associations, all of which contributes to the clarification of the domestic trade. Finney was a substantial tradesman, whose financial and trading practices exemplified all the elements of the Lancashire trade both large and small, Manchester's relations with London, and the vigorous ties of domestic trade cultivated and nurtured with western Britain.

[11] Defoe, *Tour*, i, 239.

Bodmin, Padstow, St Teath, and Liskeard. Moreover, trade did not end at the Devon border. Return journeys to Bristol included stops at Barnstable, South Molton, and Tiverton in Devon, while occasional stops were made in Dulverton on the way to Taunton in Somerset.[12] Finney's and Davenport's perambulations through the counties of western England brought the partners a healthy trade. But they also employed a chapman, R. Lomax, to carry on along the road from Bristol to Truro and back, then on to Bath, Leamington, Gloucester, and Manchester, along the cross-post road so enthusiastically described by Defoe. Besides which, Lomax made the odd detour to points such as Wincanton and Sherbourne in Dorset, plus Colne and Warrington in Lancashire. He covered a lot of country, and would certainly have helped sustain and expand the trade of these Manchester dealers. In 1731 the first round trip between Bristol and Truro took nineteen days, during which Lomax was accompanied by three laden packhorses, at the expense of five shillings daily for himself and the horses. During the return journey to Lancashire the cost was much less, suggesting that use was made of the carrier service traversing the well-travelled route from Bristol north to Manchester. Lomax needed only twenty-one shillings and sixpence to pay the transportation costs for what were certainly much lighter packs of goods, stopping in Leamington and Gloucester for his customers there. The stops after Bristol were occasional only, the bulk of the trade being further west. The irregular trade on the route home suggests a representative ready and willing to sell whenever the opportunity presented itself.[13]

Throughout the 1730s Finney and Davenport employed all available means of transportation to serve their market. Coastal vessels carried bales of Manchester wares to be disbursed from Bristol and Truro, as secondary and tertiary distribution points. Bristol was the largest westerly entrepôt for both land and sea trade, supplying the inland and the Truro warehouse.[14] The coastal route channelled manufactures to key centres, after which the land-based modes of trade and traffic took over. At this point carrier services could not move merchandise as cheaply as could the coaster captains. So see routes and country roads were used in combination to bring stock to customers throughout the west of England. To sustain these markets Finney, Davenport, and their chapman had to keep in more-or-less constant motion, undertaking the relentless circuits across the moors, along lowlands, and

[12] DFF/20/5, 9, 22, 32, Cheshire RO.   [13] DFF/20/8, Cheshire RO.
[14] DFF/20/17, 36, 39, Cheshire RO.

over hills, accompanied by pack-horses or followed by heavy bales on the local carrier's wagon. The momentum of this travel drew together the points of supply in Manchester, the regional warehousemen, and their more distant markets in the west: the common and continuing feature of this trade was the centrality of Manchester as the first source of supply.

Joseph Harper worked in the mercery trade in the middle decades of the eighteenth century, operating a shop in Hinckley, Leicestershire. Harper provides a different perspective on the textile trade as a man whose business was exclusively that of middleman, although, in common with Finney, Harper too travelled extensively. Harper required regular consignments of materials at reasonable prices for trade inside and outside his shop. A memo-book for the year 1753 indicated that fabrics purchased were both practical and fashionable, with textiles and haberdashery as well as grocery items, books, and stationery comprising part of his stock. Among the items noted under 'Drapery' were printed linens, dark-ground chintz, and light-blue chintz; while under mercery, black velvets, and 'Stript Cotton' were listed.[15] All these materials were then being manufactured in Lancashire and vicinity.[16] Thus there can be no doubt that Harper was being furnished with an assortment of British-made cottons and cotton-linens.

Harper's earlier records for the years 1733–7 illustrate the recurring round of journeys he undertook. His trade was more limited than Finney's, but travel was indispensable to both. Harper travelled for two reasons: either to sell his merchandise or to purchase stock. The regular sequence of journeys to Rugby and Atherstone, for example, indicate that he attended nearly all of the local markets and fairs. The seasonal fairs appear to have prompted many of Harper's journeys, especially in the early few years when he attended most fairs in the vicinity of Hinckley. Although Wadsworth and Mann concluded that such seasonal events declined in importance as the century progressed,[17] clearly the Coventry Fair, as well as several others, retained sufficient importance to attract Harper year after year.[18]

If selling required regular travel so too did buying. To this end

---

[15] Harper (Burton-Latimer) Collection 6/87, Northampton Record Office.
[16] G. G. 2, Livres Eschantillous ..., Musée Centrale des Arts Décoratif, Paris.
[17] Wadsworth and Mann, 238–9.
[18] Harper (Burton-Latimer) Collection 6/81, Northampton RO. See Appendix 3 for catalogue of journeys.

Joseph Harper regularly made his way to London, as too had Samuel Finney. The first two years of the ledger tell of Harper's yearly spring journeys to London, where he could select stock from among the legions of shopkeepers, warehousemen, and traders. During the last two years recorded the number of trips to London rose from one to three in 1736, plus another London trip in the winter of 1737, just prior to the ledger's end. At the same time Harper cut down on the secondary jaunts to smaller markets and fairs.[19] Harper's ledger reflects the specialization of distribution-points both wholesale and retail, whereby retailers looked to a central outlet for their merchandise. Significantly, it was at this same time that he made his first trip to Manchester; to the source of the 'Stript Cotten' and other cotton textiles noted in his memo-book. On this first trip he remained for four days. The simple entry is not proof-positive of Manchester's mercantile strength, but the appearance of this trip is certainly suggestive. Taken in conjunction with evidence both before and after Harper's venture, it illustrates the commercial magnetism exerted by the manufacturers and warehousemen concentrated in the environs of Manchester.

Joseph Harper's catalogue of travels crystallized a competition that persisted through the eighteenth century between the two distribution centres in the cotton trade, the one in London and the other in Manchester. But there is no doubt that of the two cities London was the more favoured. By mid-century, London stood at the centre of a comprehensive transportation network. Eleven per cent of the population lived in the capital, while at least one in six of the nation's adult population visited London at least once.[20] Publishers produced directories and guides to the metropolis in greater numbers after 1740 in an effort to cater to the needs of visitors. These volumes enabled tradesmen such as Harper to find the desired linen-drapers, haberdashers, and warehousemen. The title of R. Baldwin's 1753 guide indicates the audience to whom this volume was directed—*A Complete Guide to all Persons who have any Trade or Concern with the City of London*. Included was a detailed index of the inns from which carriers and coaches set off and the times and days of service. London maintained close and reasonably regular ties with all points of the kingdom. In 1753, coaches travelled to York seven times weekly in

[19] Ibid.
[20] E. A. Wrigley, 'A Simple Model of London's Importance in Changing English Society', *Past and Present*, 37 (July 1967), 45–50.

summer and six in winter; Bath boasted eight weekly services in summer and six in winter; while even ports such as Newcastle upon Tyne, which one might expect to be served exclusively by coaster, had a twice-weekly service from London.[21] The proliferation of various almanacs and directories provided the country tradesman with all the factual information necessary to assist him in a commercial expedition to London and back home again along the turnpikes, post-roads, and cart-tracks of Britain.

The system of roads which connected the cities, villages, and hamlets was considered by some to be more of an obstacle to commercial intercourse within early eighteenth-century Britain than an aid to inter-regional trade. Literary records abound with complaints about winter quagmires, narrow, broken roads, or roads awash from rivers or sea—all conspiring to confound the rapid passage of goods or people. Individual comments by contemporaries like Defoe over the years attained the status of dogma, leading many to suppose that the physical limitations of eighteenth-century roads inhibited all but the most intrepid. Over the past two decades there has been a major re-evaluation of the role of road transportation.[22] Studies of the spread of turnpiked roads confirm that improvements began when and as the traffic generated by the economic life of the region required, or when the increase in long-distance haulage placed excessive strain on local resources. The 1734 petitioners from 'the Town of Manchester, and the several Townships of Newton, Failsworth, and Oldham' knew how closely tied their prosperity was to passable roads. Textile manufacturing in the north-west and commercial links with other parts of the kingdom were jeopardized by the hazardous routes. The concern exhibited in the petition was in proportion to the risks presented to the economic life of the area. Boggy or craterous roads crippled trade.[23] One of the most significant revisions to emerge from the reappraisal of land traffic is that, poor roads notwithstanding, road carriage was frequently the preferred choice among merchants. Mud, precipitous

---

[21] R. Baldwin, *A Complete Guide to all Persons* ... (1753), 99, 114, 123.

[22] Eric Pawson, *Transport and Economy* (1977); W. Albert, *The Turnpike Road System in England, 1665–1840* (1972); G. L. Turnbull, *Traffic and Transport* (1979); id., 'Provincial Road Carrying in England in the eighteenth century', *Journal of Transport History*, NS 4 (1977); M. J. Freeman, 'Transporting methods in the British cotton industry during the Industrial Revolution', *Journal of Transport History*, 2: 1 (1980); J. A. Chartres, *Internal Trade in England, 1500–1700* (1977); T. S. Willan, *The Inland Trade* (1976), provide a selection of the reappraisals of road transportation in the domestic trade of England.

[23] *Journals of the House of Commons*, 22 (1734), 372.

hills, narrow tracks, and deeply rutted roads tried the endurance of driver and vehicle, but travel by road was still faster and more reliable than either coastal or river routes.

Water-borne shipments suffered substantially more from the vagaries of wind and weather than did goods carried by wagon or coach. In the 1730s Samuel Finney depended very heavily on coastal shipping to move his supplies from Manchester to Bristol and Cornwall. Letters were commonly exchanged overland, in which complaints or sincere hopes were voiced over the fate of sea-bound cargoes. The letter of 7 July 1732 is one such missive: 'Robert has been rady for the Sea sometime but can(?) not Sail on the tuesday last having no wind to Send him away'. Later, Finney expresses the hope that his correspondent might shortly expect the arrival of an awaited cargo, for the captain had set sail fourteen days ago, 'and most of the time till now the wind has been Costerly [*sic*]'.[24] Commerce by sea awaited favourable winds and tides, and suffered the rigours of winter storms and howling gales, such as those reported by William Stout. He described ships immobilized by ice and destroyed in harbour and on the seas during winter weather, while river traffic experienced delays and obstructions as well. Stout also complained of the damage inflicted by rats on sea-cargoes—an endemic hazard but not the least of a merchant's worries.[25] During the frequent periods of war, offshore enemy vessels made all coastal shipping problematic, putting further pressure, and greater dependence, on land routes for inland trade. By mid-century, thirteen arterial routes branching out from London were improved by turnpike trusts. These highways stretched to all corners of the kingdom from which London received supplies, knitting together provincial capitals and market regions to the metropolis. Harper noted in his ledger that 'Goods Came from London' only several days after his own return from the capital.[26] At that time the consignment from London would have travelled most of the way along the amended turnpike surfaces. Thus the meliorations in the nation's highways contributed to the vigour of domestic trade and London's strength as the principal distribution centre.

G. L. Turnbull's study of Pickford's carrier service revealed material about what is probably the oldest recorded continuing carrier company. Pickford's grew out of the demand for carriers to move the

[24] DFF/21, 7 July 1732, 28 Oct. 1735, Cheshire RO.
[25] J. D. Marshall (ed.), *The Autobiography of William Stout* (1967), 108, 227, 228.
[26] H(B–L) 81, Northampton RO.

Lancashire cottons from Manchester to London. First mention of the firm appeared in a 1756 edition of the *Manchester Mercury*, and by that time James Pickford ran a well-established undertaking.[27] Pickford and several other carrier firms operated on the Manchester–London route, funnelling the Lancashire manufactures to the nation's largest market. John Holker, when spying for France, reported that each week approximately one thousand bolts of chintz were sold by Blackburn manufacturers to London merchants for dyeing.[28] Holker's account gives an estimate of the productive strength of one section of manufacturers, leaving the larger portion of the industry to speculation. Nevertheless, the figure cited hints at the quantity of textiles moving each week in the direction of London. The momentum of trade intensified with each decade of the century; and with this commercial activity the basis of a distribution infrastructure was formed. The practical requirements of this level of road-shipment demanded temporary storage space at off-loading points along the route and at the terminus of the journey. Innkeepers fulfilled many of the secondary administrative functions throughout the whole length of the carrier networks, selling tickets, arranging transfers of packages, and providing temporary warehouse space. It was this last activity that developed into part of a national distribution network, although not without problems that inevitably dogged such expanding enterprises. Samuel Finney felt justified in his complaint over the inadequacies of the warehousemen and carriers employed to send his order to Bristol. Finney's letter of 7 July 1732 advised that further precautions be taken:

> just how rec'd advise of Mr. Cox . . . that none of the Hosewares arriv'd[.] it's an insufferable neglect that Mr. Jordan shou'd take no better Care in the dispatch of the Same and think it will be advisable to send Robin to Mr. Kellands ware House what Carriers took 'em for Bristol[.] if sent away as its probable some might and not know where to deliver 'em in Bristol.[29]

Delays notwithstanding, London channelled and re-routed an untold quantity of merchandise in the inland trade.

Various of the London inns specialized in coach-and-carrier services to selected parts of the country and it was around the terminal inns in London, where the carriers from the north-west ended their

---

[27] Turnbull, *Traffic and Transport*, 6.
[28] G.G. 2, nos. 88–100, Musée Centrale des Arts Décoratifs.
[29] DFF/21, 7 July 1732, Cheshire RO.

journey, that a warehousing system developed expressly for Manchester wares. Initially the bales of cloth were probably housed temporarily in the outbuildings of inns such as 'The Axe' on Aldermanbury, 'The Bell Inn' on Wood Street, 'Blossom's Inn' on Lawrence Lane, 'The Castle and Falcon' on Aldersgate or 'The Swann with Two Necks' on Lad Lane.[30] The profits to be made in selling these goods to wholesalers encouraged development of a separate warehousing business over the course of the eighteenth century. Even at the beginning of the century sellers of textiles, ready-made gowns, or banyans, positioned themselves in inns and coffee-houses in London. Harrison's warehouse, for example, was situated 'Against the King on Horseback, Charing Cross'.[31] Warehouses naturally arose close to the area at which the carriers arrived as a purely practical consideration. J. A. Chartres has described the process of specialization of London coaching inn, the sites at which goods and tradesmen were received and dispatched. London's inns developed as entrepôts before those of any other centre, a fact Chartres ascribes to 'rather earlier and more persistent demand for inns to serve the road transport industry'.[32] Architectural style and spacial organization of the inns enabled shops, wholesale warehouses, and associated transport trades, such as smithies and farrier shops, to congregate within their walls and yards. A 'confederacy of business interests' characterized settings like 'The Axe' or 'The Bell Savage Inn', gradually leading to a specialization in the trade and traffic from one region of the country. 'Thus the point of arrival became, in many cases, the effective market place'.[33] As the metropolitan shipments of cotton textiles increased, a district devoted to wholesale distribution developed within the area of Lad Lane, Wood Street, Lawrence Lane, and Aldermanbury, growing out of the established textile district of Cheapside. Some warehousemen used the storehouses built along the lanes and courtyards in that neighbourhood, but others remained either in inn buildings or the courtyards of those premises. All congregated in the vicinity which specialized in their trade. Manchester warehousemen Kearsley & Chesnie, for instance, were listed in a 1776 directory at '2 Bloosom's Inn Gateway,

[30] *The Universal British Directory*, i, 474–90.
[31] *The Country Journal*, 3 Aug. 1728.
[32] J. A. Chartres, 'The Capital's Provincial Eyes: London's Inns in the Early Eighteenth Century', *London Journal*, 3: 1 (1977), 25.
[33] Chartres, 32, 25–6, 32–8. S. D. Chapman describes such a process among the wholesalers of cotton stockings in 'Enterprise and Innovation in the British Hosiery Industry', *Textile History*, 5 (1974).

Lawrence Lane', while warehouseman Bloss Branwhite conducted his business from 'The Spread Eagle Inn', Gracechurch.[34]

The growth of the inland trade in the cotton industry saw the parallel growth of an interdependent group of tradesmen, warehousemen, innkeepers, and carriers, all of whose prosperity was founded on the movement of cotton textiles to markets throughout the land. The progress of bales of woven cottons from the workrooms of Lancashire, Cheshire, Derbyshire, Nottinghamshire, and Yorkshire to shelves and counters in shops around the nation, depended on the extensive carrier system in operation along most stretches of the nation's roads. The turnpike carriers like Pickford, Bass, and Twiss are well known. But in addition to these were a multitude of anonymous men and women running short-distance haulage services, like Mary Baxter who travelled only from Liverpool to Warrington, described as coming to 'Luke Carrol's in Elbow-lane, every Wednesday; returns the next morning with goods for all parts of the road'.[35] Similar services were noted for thousands of other intermediate destinations. Henry Marsh was more ambitious in the service he offered, arriving in Liverpool at Dale Street each Tuesday, Wednesday, and Saturday and returning the same day 'to Warrington, Manchester, Stockport, Macclesfield, Knutsford, Leicester, Nottingham, all parts of Derbyshire, Sheffield, and all parts of Yorkshire'.[36] In addition to making deliveries to the main towns in Lancashire, Marsh continued on with cargoes for all points east, south-east, and north-east on the map. The medley of carrier services ensured deliveries between cities and towns, between towns and villages, and even to individual houses along the route.

Along the roads of Britain an army of carriers moved mountains of goods from ships to work-rooms, from warehouses to shops, from sellers to buyers. In 1766 it was computed that '312 tons of cloth and Manchester wares' travelled from Manchester through Stafford weekly.[37] The regularity of this enormous undertaking attests to the relative efficiency of the bulk-transportation system, an essential mechanism in servicing the home market. Furthermore, it must not be overlooked that every improvement in transportation served the commercial interests of Manchester as well as the capital. Every mile of turnpiked road allowed the Manchester-based wholesaler at least an equal opportunity, if not an outright advantage, in moving products to

[34] H. Lowndes, *A London Directory* ... (1776), 21, 95.
[35] *The Universal British Directory*, iii, 668.   [36] Ibid.
[37] R. Whitworth, *The Advantage of Inland Navigation* (1766).

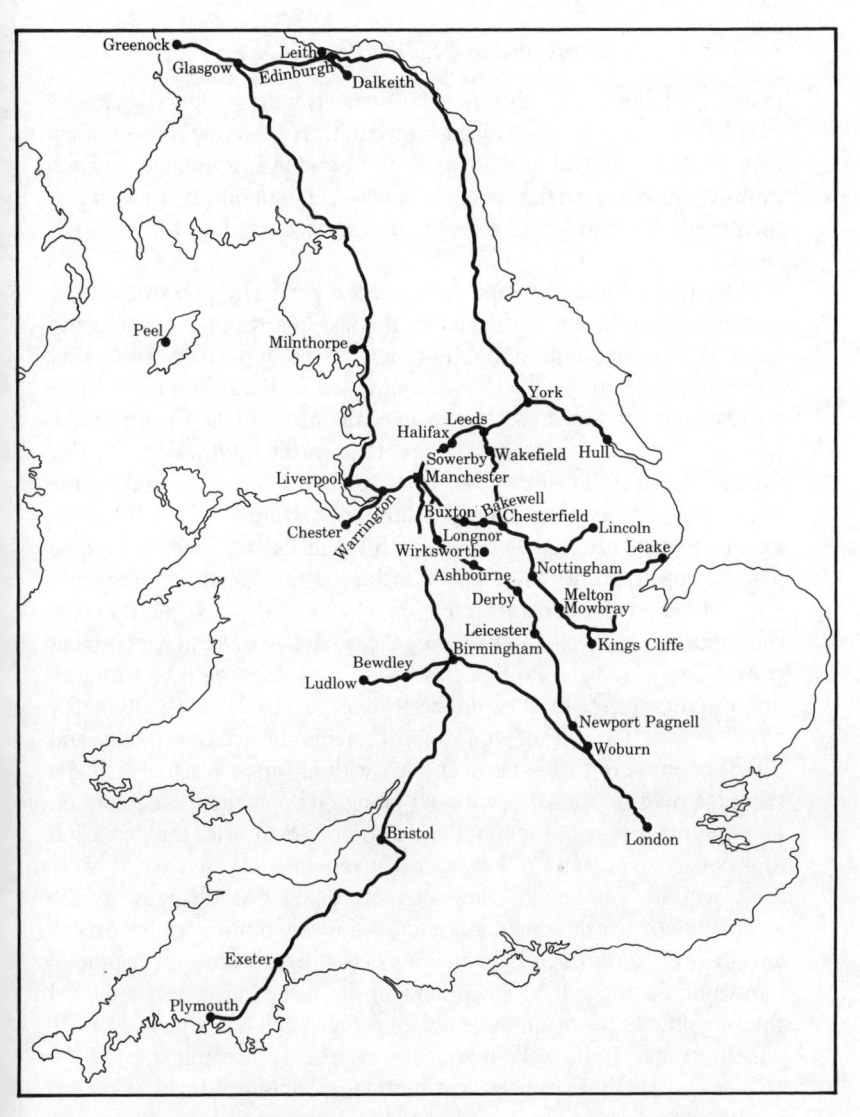

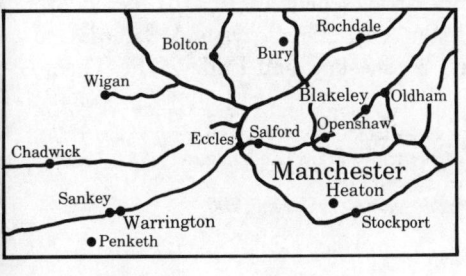

Location of customers supplied by the unknown Manchester firm 1773–79

market directly. The distance between London and Manchester enabled the smaller, specialized centre to thrive, serving its customers face to face, in addition to stocking London's warehouses. Each improvement in carrier service, every diminution in travel-time, furthered Manchester's role as a commissary for the nation's shopkeepers.

Where once a man would have seen little if any improvement in transportation during a lifetime, by the last quarter of the eighteenth century improvements and innovations were appearing with ever-increasing regularity. The services outlined in Baldwin's 1753 guide represented accumulated reforms over the previous half-century: the next ten years brought ever-more noteworthy innovations.[38] The future Viscount Torrington, an inveterate traveller, remarked on the improvements made for travellers through Oxford in 1781. 'Oxford', he wrote, 'had been lately much improv'd in its Inns (which were so justly complained of) and the new stables at the Angel are excellent'. On a later visit he commented on changes apparent in the city, recording 'a long walk ... to survey the progress of the new pavement from Mary Magdalen, to St Giles Church, which should be strong to endure the traffic of the numberless stage coaches, which hourly pass'.[39] But not all of the improvements in transportation and distribution were to his taste. 'I wish with all my heart that half the turnpike roads of the kingdom were plough'd up, which have imported London manners ... I met milkmaids on the road, with the dress and looks of grand misses.'[40] No matter how repugnant these changes were to Viscount Torrington, they marked qualitative changes in the availability of cheap, abundant, and fashionable fabrics for milkmaids and their ilk all through Britain. This diarist rightly attributed some of the blame for the spread of fashions to the turnpike-road system and the ongoing series of improvements in the coach-and-carrier service which brought in its wake new styles in dress and manners.

Newspapers during this period commonly announced schedules and additions to existing coach-and-carrier schedules. In 1767 the Oxford 'Machine' left from Oxford Tuesday, Thursday, and Saturday to arrive in London on Monday, Wednesday, and Friday. The Thame

---

[38] Baldwin, *A Complete Guide* ... ; Thomas Mortimer, *Universal Director* ... (1763). Two examples of additions to service are—Bath, 1753: 6–8 coach trips, 5 carrier trips weekly; 1763: 15–18 coach trips, 15 carrier trips weekly. Kidderminster, 1753: 1 carrier trip weekly; 1763: 5 coach and carrier trips weekly.

[39] G. B. Andrews (ed.), *The Torrington Diaries* (1936), i, 5, 206.

[40] Ibid. 6.

stage-coach left on Thursday, collecting and unloading passengers and packages at Chinnor, Bledlow, and Risborough, then lumbering on to London.[41] In the same year the 'Birmingham and Stratford Fly' advertised a one-day journey along the London-to-Birmingham route, with coaches leaving three times weekly.[42] Improvements in scheduling or in the construction of the coaches were proudly announced to the public; as in the *Manchester Mercury*, which proclaimed the arrival of a coach set 'Upon STEEL SPRINGS' for a more comfortable run into Liverpool.[43] The five-volume *Universal British Directory*, published intermittently from 1790 to 1798, was the ultimate in guides to new routes and timetables. These five large volumes matched the requirements of the nation's tradesmen for detailed factual information to ensure the safe and speedy delivery of correspondence, personnel, and merchandise. The catalogue of coach-and-carrier services to and from Manchester, for example, comprised five closely printed pages. Market demand had accelerated the rate and efficiency of distribution in the same way as it had the production of the fabrics themselves.

MIDDLEMEN IN THE COTTON TRADE

In London the sale of Manchester wares was undertaken by tradesmen according to several arrangements. M. M. Edwards has described the various specialities of these middlemen: the commission agent, for example, sold British-made cottons on commission, simultaneously keeping the producer informed of all shifts in fashion that could influence sales.[44] Warehousemen comprised another category of tradesmen, although they could be described variously as linen-dealers or given other titles that disguised their function as a middleman in trade. Of this group Samuel Salte is certainly the most well-known. George Unwin has defined the responsibilities of the warehouseman as being:

to buy from country manufacturers and from importers to sell to retailers. They had to keep in constant and sensitive touch with the fluctuating demand of fashionable consumers on the one hand and with the technical resources and business capacity of an everchanging body of manufacturers on the other. They had to be prepared to offer credit both to manufacturers and retailers. ... The great, 'Linen Houses', to which class Saltes' probably

---

[41] *Jackson's Oxford Journal*, 24 Jan. 1767.
[42] Ibid. 25 Apr. 1767.   [43] *Manchester Mercury*, 22 Feb. 1780.
[44] M. M. Edwards *The Growth of the British Cotton Trade, 1780–1815* (1967), 151–7.

belonged, are said to have needed a capital of £30,000 to £50,000 for their establishment.[45]

Thomas Mortimer's *Universal Director* for 1763 contains listings of all the major tradesmen and warehousemen resident in London, included in which were eighty wholesale linen-drapers and warehousemen, six of whom traded exclusively in Manchester wares. Thirteen years later the expansion of production and trade were reflected in the greater numbers of metropolitan middlemen. *The London Directory* for 1776 contained 130 warehousemen, of whom twenty-four traded in Manchester goods; there were in addition fifteen wholesale linen-drapers and eight wholesale mercers, making a total of 153 designated wholesale dealers in textiles. However, even this extensive collection of wholesale dealers did not constitute the entire assemblage. The title ascribed to a trader did not always indicate all business functions. S. & W. Salte, for instance, were listed as linen-drapers, giving little indication of the importance of the wholesale aspect of Saltes's occupation. In fact, a multiplicity of functions was common among tradesmen in the period. Advertisements regularly appeared placed by mercers, drapers, and the like, soliciting custom from country tradesmen.[46] Thus an appraisal of the numbers of London's retail tradesmen is also of interest since these firms were, in many instances, active middlemen. In 1763 Mortimer tallied forty-six mercers and 155 linen-drapers of sufficient stature to warrant an entry in his directory. Twenty years later London boasted over 100 of the former and nearly 280 of the latter tradesmen; a peerless collection of wholesalers and retailers in the textile trade.[47]

The function of the warehousemen and wholesale linen drapers was essentially that of the intermediary. In order to facilitate the movement of information and stock between one firm and another, the London traders specializing in cottons, linens, and the like congregated together. Cheapside and the Poultry was the backbone of the textile-trading area, the streets, lanes, courts, and alleys that ran off that main thoroughfare, north and south, containing hundreds of this company. Milk Street, Aldermanbury, Lad Lane, Lawrence Lane, King Street, Ironmonger Lane, Wood Street, Bucklersbury, and Princes Street

---

[45] George Unwin, *Samuel Oldknow and the Arkwrights* (1924), 56–7.
[46] *The Gazetteer and New Daily Advertiser*, 29 July 1779, 25 Aug. 1779; *The British Gazetteer*, 9 Mar. 1797.
[47] Mortimer, *Director*, 107–65; R. Baldwin, *The Complete New Guide* . . . (1783), 190–323.

TABLE 4.1 *London tradesmen operating as wholesale distributors*[48]

	1763	1776	1790
Warehousemen*	74	129	160
Manchester warehousemen	6	24	41
TOTAL	80	153	201

Note: * all wholesale line-drapers and mercers included.

near the Mansion House held the majority of both regular warehousemen dealing in mixed textiles and the Manchester warehouses, specializing in cottons. In and around the same area gathered the linen-drapers, wholesale linen-drapers, and a range of subsidiary and affiliated trades. (See Plate 18, Greenfields' warehouse dealing in multiple textile products.) By the last decade in the eighteenth century the number of warehouses dealing exclusively in cotton textiles had risen to forty-one, an increase of thirty-five since the first listing of that trade in 1763. London's advantage lay in the magnitude of its commercial activities, transportation facilities, and market-size; but above all London had the cachet of fashionability. For provincial tradesmen the announcement that stock had recently arrived from London was a guarantee of added custom. This boast can be found in advertisements around the nation, and in itself hints at the vast stocks moving in and out of the capital.[49]

London has been described as an engine of economic growth. The concentration of wealth and the sheer size of the market generated a unique demand during the sixteenth and seventeenth centuries. By the second half of the eighteenth century prosperity also accrued to the capital's wholesalers, supplying goods required by the rest of Britain. Metropolitan middlemen dominated many trades. But did London's commercial might crush all competing distributive centres? Did the colossus of trade dominate the distribution of British-made cottons? In most instances metropolitan pull and the concentration of service and supply checked any contesting regional sites; the commercial pre-eminence of the capital remained absolute. S. D. Chapman traced the

---

[48] Mortimer, *Director*, 107–65; Lowndes, *Directory*, 4–194; *The Universal British Directory*, i, 49–345.

[49] *Jackson's Oxford Journal*, 2 June 1770, 14 Feb. 1776; *The Bath Journal*, 24 Sept. 1753.

dominance of London as the distribution-hub for knitwear from the counties of Nottingham and Derby. These provinces were traditional knitting centres that received a boost with the establishment of cotton mills in their vicinity. Gradually much of the stock began to be made from cotton yarn, superseding linen, silk, and wool. Between 1770 and 1775 Chapman discovered nearly seventy manufacturers from those regions with identifiable ties with London. Some marketed their hose through a commission agent, others maintained a warehouse in the Wood Street–Lawrence Lane district off Cheapside. Some of this number even kept up a house in London to facilitate the London side of the business. But in all cases London remained the unrivalled distribution centre; no competitive provincial site arose to challenge London in this capacity.[50]

Manufacturers of checks, chintzes, and fustians had taken a hand in the sale of these products since the inception of the trade. Initially accompanied by heavily laden pack-horses, the Manchester Men brought their stock-in-trade to the shopkeepers, fairs, markets, and households along the routes that criss-crossed the country. In 1704, these specialized traders were recognized for their vital role in 'extending the home consumption of cotton goods, especially for women's wear', and were exempted from the new tax on common hawkers.[51] The string of pack-horses trekking across the landscape was a common sight that continued into the second half of the century. However, as the fairs declined shops proliferated[52] and fixed retail outlets expanded the demand for cottons in a way that could not be efficiently met by the old habits of distribution of the Manchester Men. During this time alternative distribution systems were developing, ultimately supplanting the slow and cumbersome circuits of the Manchester Men. But Manchester's capacity as a distributor was not superseded. Manufacturers of cotton goods retained their marketing capability. This phenomenon has been well documented, one of earliest of these enterprises being J. & N. Philips & Company, manufacturers of smallwares and tapes. The records of the Philips Company provide another illustration of the marketing ventures undertaken by the manufacturers themselves. Wadsworth and Mann were the first historians to identify this further development of the inland trade. From the first the Manchester manufacturers sold their

[50] Chapman, 'Enterprise and Innovation', 33–5.
[51] Wadsworth and Mann, 46, 238.
[52] T. S. Willan, *The Inland Trade* (1976), 88.

products in the face of opposition emanating from the capital's middlemen, who sought to monopolize the wholesale trade.

The letter-book for the years 1753 to 1769 is the earliest surviving remnant of the Philips' trade. John and Nathaniel Philips began their career in 1747, specializing in smallwares and tapes—an essential component in the making of clothes.[53] The success of this firm depended upon broad markets beyond the immediate confines of Lancashire. Thus the partners conducted wide-ranging expeditions through many parts of Britain, calling on shopkeepers and soliciting sales for their merchandise. Their circuit took them through the midlands and East Anglia, in addition to more westerly trips to Bristol and north into Scotland. The collection of bills for the year 1756 locates customers in towns such as Bristol, Burton, Wolverhampton, Nottingham, Cambridge, Sandy, and Manchester.[54] Aside from direct sales Philips sold their tapes and smallwares to London warehousemen, and it was these tradesmen who revealed a strong disinclination to compete for customers with the manufacturers. The London middlemen were not satisfied to receive a portion of Philips' products; they wanted the complete output. Several pressing requests to that effect were made to Philips by London warehousemen—requests always refused. London warehousemen did not happily accept competition, and appeared to want a strict delineation of function with themselves as distributors and Philips restricted to production. Regardless of the grumbling of the London tradesmen, cotton manufacturers persisted in acting as middlemen. One metropolitan warehouseman complained that he was being undercut by the Philips' trade; a charge refuted by Philips: 'You don't do us justice in charging us with selling small Quantities to Haberdashers as cheap as to you, and we assure you we never did'.[55] While Philips was happy to smooth relations to this extent, he was not prepared to go further. Producers remained integrally involved in the identification and cultivation of markets and in the retail and wholesale distribution of fabrics. The productive capacity of the Manchester region grew in tandem with its distributive strength: Manchester-based wholesale distributors rose from 50 in

---

[53] M 97 Je N.P., J. & N. Philips & Company Records, Manchester Public Library.
[54] M 97/3, Manchester Public Library.
[55] Ibid.
[56] Elizabeth Raffald, *The Manchester Directory for the Year 1773* (repr. 1886), 5–52; *Lewis's Directory for the Town of Manchester and Salford for the Year 1788* (repr. 1888), 5–33; *The Universal British Directory*, iii, 789–853.

1773 to 60 in 1788, and their number more than doubled to 141 by 1794.

Manchester could not match the size of London's diverse wholesale network. However, where London tradesmen dealt with all manner of merchandise from domestic and foreign sources, Manchester's warehousemen concentrated on a specific selection of mainly cotton commodities. When considered in this light, the strength and capacity of the Manchester-based system becomes more apparent. This industrial centre directed a challenge at London's commercial hegemony and succeeded in its aim, retaining a strong hold on the wholesale dispersal of British-made cottons. The numbers of middlemen noted above suggest a part of this commercial strength, but in addition, individual manufacturers like Finney, Davenport, and Philips, undertook the sale of their own merchandise. With each passing year Manchester assumed a greater stature as the key entrepôt for the cotton industry.

Manufacturers in smaller outlying towns and villages were obliged to retain warehouse space in Manchester as more buyers were attracted annually. Glasgow manufacturers also kept warehouses in Manchester, recognizing it as the capital of the industry. In 1773 over a hundred fustian manufacturers and twenty-four check manufacturers from the neighbouring manufacturing towns of Lancashire, Cheshire, and Derbyshire arranged for warehouse-space in Manchester to display their goods. By 1781 this number had grown. Approximately 170 fustian manufacturers and twenty-eight check manufacturers were represented, in addition to local producers.[57] Of all the many cotton towns, only Macclesfield maintained a warehouseman.[58]

Manchester's wholesale and retail distribution flourished, in spite of the resentment of London's wholesale dealers. London dealers, such as Samuel Salte, continued to demand exclusive rights of distribution for choice cotton textiles, but this monopoly was always denied them. Samuel Oldknow refused Salte's entreaties, as Philips his precursor had done, retaining his independence in the marketing of his products. Oldknow sold his goods widely; a list of customers between 1782 and 1786 yields nearly 110 names from twenty-nine destinations, including those in trade and private purchases.[59] The united strength of wholesale dealers and manufacturers held off the commercial challenge

[57] Wadsworth and Mann, 255–60.
[58] *The Universal British Directory*, iii, 897.
[59] Oldknow Papers, 773, John Rylands Library, Manchester.

of the capital. Manchester's active role in the dispersal of its products remains one of the distinctive characteristics of the eighteenth-century British cotton industry.

MOVING GOODS TO MARKET: PEDLARS, SHOPKEEPERS, AND A MANCHESTER FIRM

Lancashire's manufacturers found it more efficient to relinquish the long journeys with laden horses; thus the peripatetic Manchester Men were gradually superseded by manufacturers or their agents travelling with samples only, visiting shops large and small around the country to obtain orders. But this did not ring the death-knell for all the itinerant pedlars and hawkers supplied from Manchester. Despite the slow but steady growth of retail outlets throughout the nation the hawkers, pedlars, and travelling Scotchmen remained a significant source of merchandise for tens of thousands of Britons in rural settings distant from town and village shops. The hawker brought the fashions of the town to the country, albeit a little late. The roving pedlar formed an essential link between the manufacturers of fashionable low-cost fabrics and the vast, diffuse market outside the towns and villages. Pedlars continued to ply their trade throughout the country, undiminished in their efforts in the face of the great transformations taking place in the number and sophistication of the urban shops. The country vicar James Woodforde, residing in the East Anglian countryside, noted in his diary the arrival, in May 1778, of Hannah Snell at the local hostelry. The woman had served in disguise as an infantryman and now used her notoriety to solicit custom for her modest trade in hosiery and haberdashery wares.[60] Snell was only one of several regular callers in this part of eastern England, for the paths of the pedlars and hawkers intersected the provinces. Woodforde later wrote of another pedlar who called in November 1781, a Mr Alldridge, 'who goes about with a Cart with Linens, Cottons, Lace etc.'. This more-affluent pedlar received a bonus when he called on the clergyman, for Woodforde was entertaining guests at the time and they too were ready to buy from the attractive goods presented them. Woodforde noted that:

I bought of him some Cotton 6 Yrds for a morning gown for myself at 2/6 per yard, pd. 0.15.0 [.] Some Chintz for a gown for Nancy 5 yds and ½[,] I pd,

[60] James Woodforde, *The Diary of a Country Parson, 1758–1802*, ed. John Beresford (1924), i, 224–5.

1.14.0 ... Nancy also bought a Linen Handkerchief etc. of him. Mrs. Howe bought a silk handkerchief of him also.[61]

The extent and utility of the pedlar's trade even as late as the last decades of the eighteenth century can be judged by the response of the cotton manufacturers to the threatened legislation designed to abolish the migrant traders. In 1785 a torrent of protests poured from the manufacturing centres of Britain. In the petitions to Parliament from 'merchants and wholesale dealers of Liverpool' and 'Linen Committee, Silk Manufacturers and Callico Printers of Glasgow', the value of the sales dependent on the pedlars was asserted: 'Sale, which is generally from House to House in Country Villages and Districts, remote from Towns where Shopkeepers reside'. The petitioners insisted that without the services of the hawkers and pedlars 'great Quantities of *British* Manufactures' which were now sold would not find a market.[62] In those counties in the north where settlements were scarce and retail outlets scarcer, the householders depended almost entirely on the fabrics and haberdasheries brought them by the pedlars. The pedlars' trade differed from that of the settled mercer or draper, not only in the wide geographic region served, but also in the provision of goods on long-term credit, in consideration of the seasonal income of the land-based customers. The activities of the pedlars complemented those of fixed retailers, asserted the petitioners. A group of Kendal pedlars explained how they always 'have Stocks of Goods on Hand, and ... give Credit with the greatest Part of their Goods to a very great Number of Persons at considerable Distance from each other'.[63] The pedlars and their champions in the manufacturing trades insisted that these peripatetic tradesmen served a profitable domestic market that would be left wanting in their absence. The petition of the Society of Travelling Scotchmen, from Bridgenorth, Shropshire, provides additional evidence of the health of these retailers. Michael Macmichael, one of the principal traders of that society, had £5,000 capital invested in stock-in-trade, while debts amounting to £3,000 were owed him by his many customers. Although not every member of the society possessed an equivalent investment, most had over £1,000 committed to the trade. All of this testified to the magnitude of this part of the domestic trade, supplied directly from Manchester and other cotton-

[61] Woodforde, i, 332.
[62] *Journals of the House of Commons*, vol. 40, pp. 1,039–40.
[63] Ibid. 1,018.

manufacturing communities with 'Silk, Cotton, Linen, and Worsted, and ... almost every other Article of Female Attire'.[64]

Several Manchester warehousemen formally combined that function with work as chapmen, complementing wholesale distribution with itinerant retail vending.[65] The importance of the travelling tradesmen to the British cotton industry is further suggested by the juxtaposition of manufacturers and hawkers in the smaller mill-towns. In Stockport approximately seventy cotton manufacturers were settled by the end of the century, producing checks, fustians, and some muslins. While there was no associated concentration of warehousemen, Stockport and several other towns had an abundance of traders designated 'huckster' in the directory. It was no coincidence that nearly twenty firms and individual hucksters settled in the vicinity of the cotton town of Stockport, or that Newton in the Willows, in the neighbourhood of Wigan and Warrington, was the base for four firms of travelling pedlars.[66] Clearly the proliferation of the 'hucksters' resulted from the vigour of the local cotton manufactories in Stockport, Newton, and Blackburn, reflecting the commercial specialization of regions outside the immediate environs of Manchester.[67] The ancient ways of the pedlar had been modified with the expansion of the cotton industry and remained an essential component of the distributive network. Towns like Stockport, as well as Manchester, thrived by the provision of stock to itinerant sellers large and small. The north-west flourished in its many-sided role as distributor for the cotton industry.

The proliferation of permanent shops and stores in towns and villages of even modest size marked a qualitative advance in domestic trade and confirmed the national preoccupation with material goods.[68] Too few historians, with the notable exception of T. S. Willan, have turned their attention to the legion of eighteenth-century shopkeepers,

---

[64] Ibid. 1,017–8.   [65] *Manchester Directory*, 12.
[66] *The Universal British Directory*, v, 143.
[67] Ibid., iv, 477–82, v, 18, 143.
[68] The provincial shop with all manner of wares provided a retail outlet which by its very nature facilitated and even encouraged the consumption of goods, of which British-made cottons were a part. B. A. Holderness maintains that, although the ' "retailing Revolution" is rightly located in time no earlier than about 1860 ... the mercer's shop, dealing almost exclusively in ready-made goods, had made its appearance in country towns and in some villages at least two centuries before' (B. A. Holderness, *Pre-Industrial England* (1976), 139). A preoccupation with nineteenth-century retail expansion has often been to the detriment of a clear appreciation of the capacity of the eighteenth-century forerunners to supply the domestic market with a swelling tide of new commodities.

pioneers in the new retailing practices.[69] Part of the reason for this omission undoubtedly lies in the dearth of documentation from this section of society. None the less, such shopkeepers' records as remain are crucial for this study in order to make an assessment of both the uniformity of merchandise provided by the tradesmen and, if possible, the sources of their supplies. Fortunately, several collections of records of this type have come to light. The abundant cache documenting the trade of the Kirby Stephen shopkeepers Abraham Dent, first used by T. S. Willan, provides a continuing fund of information about tradesmen during the third quarter of the eighteenth century. The characteristics of Dent's suppliers say a great deal about the commercial life of Britain at this time. It is significant that Dent was served by approximately 190 individual suppliers, not by several general tradesmen. Those who sold merchandise directly to Dent reflect the high degree of mercantile activity in the domestic market, an activity not restricted to metropolitan areas. Willan noted all the disadvantages of the small village in which Dent resided: 'an inland town with no water communication at all; it was not near to a port, not even to Westmorland's only port, Milnthorpe; it was not very near to a large town and was itself much too small to support wholesalers of its own, except in stockings'.[70] Yet even in this unpromising environment Dent attracted the custom of traders from around the country. Regional dealers provided a range of clothing items; Kendal tradesmen, for example, sold Dent lace, bombazine, and callamanco, silk handkerchiefs, thread, and mitts. Penrith in Cumberland contained three suppliers at least who sold Dent checks, linen, and sacking. Dent was also supplied by twelve London tradesmen and approximately twelve from Manchester.[71]

As far as can be determined, three of these London tradesmen sold Dent textiles or textile products. Mr Samuel Dales, for example, carried on a trade with Dent for at least four years, from 1762 to 1766;[72] Dales was listed in Mortimer's guide to London of 1763 as a

[69] There is another happy exception to this case. The publication of *The Diary of Thomas Turner* (1985), edited by David Vaisey, marks the arrival of a fascinating account of a country shopkeeper in mid-eighteenth-century Britain. Due weight has been given to his many commercial relations and shopkeeping activities in this volume, unlike earlier published extracts from this diary, which have focused mainly on the 'curiosities' of village life.
[70] T. S. Willan, *Abraham Dent, an Eighteenth-Century Shopkeeper* (1970), 74.
[71] Ibid. 29–41.
[72] WBD/63/2, Records of Abraham Dent, Cumbria Record Office, Kendal.

linen-draper, residing in Cheapside.[73] Prominent among the London suppliers was Thomas Elton & Company, a haberdasher on Milk Street, who served Dent from 1769 to 1774.[74] The final traders of this group were the Messrs Maltby, whose record of trade with Dent covers only 1768, during which time they provided worsted and woollen goods. They were listed as Norwich factors in a London guide and operated a warehouse on Queen Street, Cheapside.[75] Detailed invoices do not survive from Dent's retail suppliers, although in some instances it is possible to determine some characterstics of the business conducted. Elton sold a substantial amount to Dent, amounting to £256. 10s. 3d. for the year 1769. This appears to have been the average volume of sales for all the various bits and pieces provided by haberdashers. In 1773, for example, Dent purchased £290. 10s. 2d. for merchandise from Thomas Elton. Samuel Dales, in contrast, sold little to Dent: £6. 15s. 2¾d. for 1762 and £4. 3s. 3¾d. for 1763, while Thomas and George Maltby made but one sale to Abraham Dent of £9. 3s. 0d.[76] The conclusion that can be drawn from these records is that Dent was prepared to try several London wholesalers, for reasons which at this juncture can only be speculated upon. Perhaps he sought to try new products, or was tempted by cheaper lines of goods. Overall, Dent retained a long-term business relationship with only one of the London textile suppliers, and that firm did not deal in cotton yard goods, but rather in the caps, hats, tapes, ribbons, and smallwares stocked by haberdashers. It is doubtful that the London suppliers provided Dent with any of the many sorts of checks, cottons, or fustian materials sold from his shop: Dent looked almost exclusively to Manchester for British-made cottons.

From 1756 to 1774 at least seven companies of tradesmen, identified by Dent as Manchester-based, sold merchandise to this shopkeeper. Of this number, two were not listed in the 1773 Manchester directory. In the case of Samuel Miller & Son, small amounts of goods were sold to Dent over periods of three years, 1756 to 1759. Flittcroft made only two sales, in January and June of 1767. The absence of these men from the Manchester directory of tradesmen and the nature of their sales-patterns suggests that they were chapmen or part of a firm of pedlars, supplied from Manchester, making long, meandering journeys through the north of England. These men carried items required by Dent and he took advantage of

[73] Mortimer, *Director*, 119. [74] Ibid. 122.
[75] Lowndes, *London Directory*, 108. [76] WDB/63/2, Cumbria RO.

their arrival in Kirby Stephen to supplement his inventory. Samuel Miller passed through Dent's village each June and February for three years.[77] The final payment made by Dent was received by a representative for Miller & Son, confirmation that this was a firm of travelling Scotchmen or pedlars, rather than lone chapmen. Although these small traders varied their routines, the organization of their trade brought Manchester goods to shopkeepers all along their routes.

The remaining tradesmen who kept Dent's shop stocked with striped cotton, chintz, silks, and thicksets were all listed in *The Manchester Directory for 1773*, or other directories of the period. All displayed common characteristics in the organization and execution of their trade. Dent's regular suppliers of Manchester wares included four well-known manufacturers: George Webster was listed as a cotton merchant; John Lawrence was described as a check manufacturer; Richard Mather as a silk and fustian manufacturer; the Messrs Bartons were listed as fustian manufacturers, as too was Joseph Gough.[78] None of the men was officially described as a warehouseman, so on the surface they were unconnected with the sale of their products. Yet each was in fact directly involved in the sale and dissemination of their products across the country, as their business with Dent indicated. Everyone of the tradesmen who arrived at Dent's shop was not a partner in the company; although some, like Henry, Richard, and George Barton did share the company business and the responsibilities. Richard Mather and Joshua Manby were also most probably partners. However, both John Lawrence and Joseph Gough employed representatives like John Lightfoot and Thomas Crompton, who appeared intermittently in the villages of Westmorland, drumming-up business and accepting payments for their respective employers.[79]

Joseph Gough sold little to Dent, small orders over several years. George Webster also conducted a modest, periodic trade with Abraham Dent from 1759 onwards. However, John Lawrence, the Bartons, and Richard Mather all carried on lengthy and extensive trade with Dent, each selling his own particular stock.[80] The credit-ledger of sales from Dent's shop includes examples of this merchandise, for instance, everlasting, thickset, linen check, striped cotton, fustian, velvet, and cotton check, assortments of fabric for which Manchester

---

[77] WBD/63/2, Cumbria RO.
[78] *Manchester Directory . . . 1773*, 8, 22, 31, 34; *Lewis's Directory for the Town of Manchester and Salford for the year 1788* (repr. 1888), 31.
[79] WBD/63/2, Cumbria RO.   [80] Ibid.

and the neighbouring region were renowned.[81] The sums expended by Dent for these fabrics were not enormous. In the nearly twenty-year relationship with the Bartons, from 1757 to 1774, Dent purchased approximately £275 worth of textiles. Richard Mather and Joshua Manby likewise provided Dent with cotton fabrics for almost ten years, leaving a record of purchases between 1765 and 1774. During that time comparatively more goods were bought from Mather, to the value of approximately £230.[82] All of the above-mentioned manufacturers utilized the then-new system of distribution whereby samples were shown and orders then taken by either the owners or their representatives.

The Messrs Barton took turns bringing items for Dent's consideration, as first Henry, then Richard, then George arrived in Kirby Stephen to display their goods. A division of responsibility in the distribution process meant that the manufacturer was no longer concerned with the carriage of orders to his customers. For that he depended on the carriers in what could at times be described as 'a very complicated system of transport'.[83] Should shipments go wrong, then it was to the carriers that the manufacturers directed their customers for compensation.[84] In this way the manufacturer was left free to concentrate on the production of fabrics and their sale to customers throughout the country.

Unencumbered journeys were quicker, the territory covered broader, and the shopkeepers served more efficiently. Sample-cards were used to display the array of colours and patterns available from the various producers, an innovation ideally suited to the selection of fabrics, however distant one might be from the source.[85] Abraham Dent was a tradesman typical of his period, stocking a wide and general range of merchandise and retaining vigorous commercial links with many areas

---

[81] WBD/63/3a, Cumbria RO.   [82] WBD/63/2, Cumbria RO.
[83] Willan, *Abraham Dent*, 43.
[84] The Philips Company received just such a complaint from one of their customers about an order that was damaged in transit. Philips responded that, 'we are sorry for the damage of the Goods—you know we have nthg to do with the Goods after they ar delvd to the Carr—you look upon the Carrier for any loss'. M97/3, Letter, Sept. 1754, Manchester Public Library.
[85] An American trader sent a pattern-card to his partner in America, writing him that, 'I have sent you a paper of patterns from Nash, Eddowes & Martin ... When there are anyone likes a pattern, send the number and say what species of goods it is and I can sent it to you'. The entire process of ordering was greatly simplified with the use of pattern-cards, and could operate from any distance with a minimum of difficulty and confusion. J. M. Price (ed.), *Joshua Johnson's Letterbook, 1771–4* (London Record Society, 1979), 8.

of the country. In spite of Dent's connection with the London commercial world, he did not choose to buy cotton textiles there. Dent chose instead to buy stock from the manufacturing centre, to be served by the energetic and efficient agents of the cotton industry.

The same preference for Manchester-based distributors was shown by a mid-century mercer, Thomas Turner of East Hoathly, Sussex. The proximity of London did not quell the mercantile zeal of the Manchester representatives, and Thomas Turner noted in his diary the pleasurable intermingling of social and commercial endeavours whenever Samuel Ridings or his agent were in the neighbourhood. On these occasions Turner arranged to walk or ride the seven miles to Lewes, where the travellers put up. There, when time allowed, Turner would meet several other acquaintances along with the travellers, enjoying a convivial evening such as the one in March 1764:

I arrived [in Lewes] about 5.20. I drank tea with my very worthy acquaintance Mr. Madgwick, and he, myself and Mr. Richardson, Mr. Tho. Woodgate and Mr. John Ridings and Mr. Fletcher spend the evening together at *The White Horse* till about 10 o'clock. We supped at *The White Horse* on some Welsh rabbits.[86]

After which—business. Following this evening of pleasantries Turner wrote: 'In the morn I arose and went up to *The White Horse* where I bought a pretty large parcel of goods of Mr. Ridings and Mr. Fletcher [Samuel Riding's servant] for myself and brother. I paid Mr. Ridings in cash £3. 4. 0*d*. in full on my account with his father Mr. Sam Riding'.[87] A similar notation appeared for a meeting that took place earlier in February of 1762. On another Tuesday morning Thomas Turner rode to Lewes 'to meet Mr. Stephen Fletcher, servant to Mr. Sam Ridings, in order to buy some Manchester goods. I breakfasted at *The White Horse* in company with my brother'.[88]

Records of Turner's dependence on Manchester-based suppliers go back to 1755, almost the beginning of the diary and continue through to March 1764, about a year before its end. London wholesalers were available and known to Turner, who made routine buying-trips to the capital. Nevertheless, he continued to purchase stocks from Ridings or his agents on their tours through Sussex. Ridings was an integral part of the local commercial community in his character as peripatetic trader, and could depend upon regular sales to

---

[86] Turner, 287. [87] Ibid. [88] Ibid, 245.

the shopkeepers in and around the south-east and in turn accepted bills drawn to their account—such was the payment Turner offered for his debt of £14. 19s.[89] Regularity of supply appeared assured, as each spring saw Ridings arrive in that region of Sussex. Occasionally inclement weather hampered the meetings between supplier and shopkeeper, but these interruptions were of short duration. On 3 March 1756 Turner noted: 'This day I appointed to go to Lewes to meet Mr. Step. Fletcher, but did not go on account that I wanted but a small parcel, and the roads very bad; so that my trouble and expense would have been more than the profits arising from what I should have bought'.[90] But 11 May saw Turner once again in Lewes concluding business with Ridings over breakfast in 'The White Horse', in the company of two other shopkeepers of the town.[91]

John Thomas of Hinckley, Leicestershire, had a more-specialized dry-goods and drapery trade for which records survive from the 1780s. A collection of sixty invoices remains for the years from 1782 to 1787. Thomas had twenty-four suppliers, and only one of these firms, McKeand and McGauchin, came from outside Manchester—in this case they were Glasgow manufacturers with a warehouse in Manchester.[92] Thomas bought a broad range of items, from muslins to Genoa corded velveret, cotton lining to jean, as well as handkerchiefs and table cloths, shawls and muslin petticoats. The London wholesalers were at a disadvantage when competing with those in such close proximity to the manufacturers. This is particularly true of shopkeepers like John Thomas who lived only a short journey from the heart of the cotton industry. Thus it is hardly surprising that the stock he required should be provided directly from the manufacturers. One cannot explain the preference of those at a greater distance so easily. There is no indication that price-variables played a part, although they may have done so. More probably the attraction to and dependence on Manchester representatives was not simply a function of possible financial advantage, but had everything to do with the greater responsiveness of manufacturer/wholesalers, situated in the middle of the manufacturing district. Ease and speed in filling orders, facility at placing special orders, and a ready adaptation to changing fashions more readily accounts for the unique distributive arrangements

---

[89] Ibid. 39–40.   [90] Ibid. 33.   [91] Ibid. 39, 345–6.
[92] Eng. MSS 1192, invoices from the papers of Mr John Thomas & Co., Hinckley, Leicestershire, John Rylands Library, Manchester.

sustained between Manchester-based representatives and provincial shopkeepers.

The irregular and intermittent nature of the existing invoices makes it impossible to calculate the full extent of Thomas's trade. However, his business does appear to be quite representative of the specialized shopkeeper in provincial England and as such is a valuable example from which to gauge the national distribution of cottons. Thomas's shop appeared to be thriving. From the steady rate at which orders were filled the demand for these cottons seemed as strong in Hinckley as elsewhere in the country. Fashionably patterned printed chintzes were sold to Thomas by prominent firms such as Robert Peel & Company; Livesey, Hargreaves & Company; Peel, Yates, Tipping & Halliwell; and T. Usher & Company. On at least one occasion, however, apologies were sent Thomas along with the invoice. One order placed with the travelling representative did not match the goods eventually sent. Livesey, Hargreaves & Company wrote in January, 1786, that: 'We are sorry it is not in our power to send the Cho[colate] Strip Chtz Patches—Those Patterns we believed at the Warehouse during the last Journey were to be worked 5/4—we lately found on our coming into the Warehouse that they have all been worked in 4/4 only'.[93] Fortunately apologies such as these were not much in evidence; one can assume that shopkeepers generally received what they ordered. Representatives and owners punctually arrived at Mr Thomas's shop, some by appointment and others on a recurring circuit. Jonathan Haworth & Sons of Canon Street, Manchester, affirmed that, 'Our Travellers will wait upon you as usual; and, be assured, that every possible care and attention shall be paid to your Interests'.[94] The clerk of cotton manufacturers John and Thomas Rideout noted at the close of one invoice that 'Mr. John Rideout will be at Hinckley in Three Weeks Time when our Order will be esteemed a favour'.[95] Such systematic and routine canvassing of shopkeepers like Dent, Turner, and Thomas speeded up the diffusion of new products throughout the country. Orders might not be perfect every time, but the basic workings of the system produced a greater efficiency in the distribution of manufactures. The 1788 trade-card of Leicester draper, mercer, and haberdasher Thomas Lomas testifies to

[93] Eng. MSS 1192, invoices, Livesey, Hargreaves & Company, Manchester 27 June 1786, John Rylands Library.
[94] Ibid., invoice Jonathan Haworth & Sons, 21 Jan. 1785.
[95] Ibid., invoice John & Thomas Rideout, Manchester, 22 Jan. 1786.

the efficacy of the distributive system. An impressive catalogue of cotton textiles and ready-made 'Manchester gowns' is included on this sheet. Within weeks or months of a product becoming available to the London public the same fabric would be stocked by ambitious shopkeepers in the provinces.

Thomas was a shopkeeper centrally located in Britain and on the route of major turnpikes, and as such he received cottons in as wide an assortment as he desired. Muslinet, printed velvet, and tape-striped muslin were all stocked, as well as the sturdier fabrics for working-men's clothes. A test of the distributive capacity of Britain's cotton trade would be to compare this collection with the fabrics in store in a more peripheral area of Britain. This can be done using the extensive draper's ledger of sales of John and Mary Morgan, of Neath, South Wales. The ledger contains a few entries for the late 1780s, but the bulk of the ledger is devoted to the purchases of their customers through the 1790s and into the nineteenth century. This volume is the only surviving one of a series, and begins several years after the close of the Thomas records. The Morgan draper's ledger quite naturally contains items that were not listed in Thomas's invoices: for example, more ready-made articles were sold by Morgan than were listed in the Thomas inventories. Several sorts of ready-made gowns, such as the 'Super Chintz Gown' and the plain 'cotton gown' were available from Morgan, plus waistcoat shapes in several sorts of fabrics.[96] All of these items were familiar from advertisements in London and other newspapers. There is no indication that Thomas had available to him in Hinckley types of textiles that were not also available for sale in South Wales. The common fustians and checks were ever-present, as were printed calicoes and furniture-checks. The various sorts of corded cloths were not generally specified by Morgan as they were in the Thomas invoices. Morgan did not take the time to identify each sort of corduroy as he listed the sale, and corduroys changed over the years. A 'Nelson's Cord' was purchased by the smith in Neath, as was plain corduroy. The former textiles celebrated the victories of Britain's favourite admiral, and not surprisingly was not available in the 1780s. 'Wild boar Tammy' was another especially popular fustian in South Wales, a district that perhaps had a greater requirement for hard-wearing materials than had Hinckley.[97] Morgan provided his town and country customers with all the varieties of fabrics and ready-made

[96] D/D ma 139, Morgan Draper's Ledger, 66, 87, 109, 110, Glamorgan Record Office, Cardiff.   [97] Ibid. 165, 166, 255, 334.

articles popular in other parts of Britain; all these fabrics—the Marseilles quilting, Japan muslin, printed calico, furniture-check, bed-ticking, Genoa velvet, Wigan stripe, and plan and printed cottons—were the product of the British cotton industry.[98] Unfortunately, no records remain that would provide a clue as to the suppliers. The point of origin of the textiles, however, cannot be disputed. Perhaps these Welsh drapers received the cottons through a Bristol intermediary or from a Manchester-based traveller. But without doubt this relatively remote area was fully a part of the national market, stocked with as broad an assortment of textiles as any other district of the kingdom.

One of the most valuable documentary sources illustrating the breadth and variety of Manchester distribution can be found in the Day Book of sales of an anonymous Manchester firm.[99] The volume details the business activities of a Manchester manufacturer who also engaged in retail and wholesale trade. The Day Book runs from 1773 to 1779, listing the divergent individuals, shopkeepers, and tradesmen who did business with this firm. From these accounts one can also determine the geographical area served through wholesale and retail sales, plus the rate at which shipments were dispatched. The unidentified Manchester firm sold to hundreds of people of all different classes, from many areas of the country, both trade and non-trade customers. The entry 'Ready Money' occurs regularly in the ledger as well, indicating that the firm acted as a retailer in addition to selling in bulk to other middlemen and shopkeepers. Cash sales tended to be for small amounts. One of the most popular articles was the ready-made gown.[100] These sold for between eight shillings to twenty-four shillings, for either a cotton or moree gown. On occasion, however, the heading 'Ready Money' hides the identity of one of the larger trade customers who paid cash for goods urgently required. A case in point was the case sale on 29 June, 1776 of twenty-two-dozen silk-and-muslin handkerchiefs, one of the first times this article had been noted. Conceivably this represented the efforts of a keen retailer anxious to beat out the competition by stocking the latest sort of accessory.[101]

---

[98] D/D ma 139, Morgan Draper's Ledger, 67, 166, 255, 334.
[99] MC: MS ff 657 D43, Day Book of sales of an unknown Manchester firm, 1773–9, Manchester Public Library.
[100] Ibid. 327, 346, 359, 364.   [101] Ibid. 351.

Anonymous cash sales were the exception. Frequently the name, location, and occupation of the buyer were recorded. Periodically, even without these clues, it is possible to discover the identity of the customers with the aid of directories. The day-book, supplemented by contemporary indexes, yields extensive and incisive information on the distribution of a whole range of textile items made in and around Lancashire, balancing the study of national distribution with an appraisal of a large multi-functional Manchester distributor. One feature of interest in the retail function of this firm was the multitude of individual customers served, those who bought textiles from this company for their own needs. A nobleman or his agent purchased several ready-made articles of clothing as well as a large piece of silk; while an overseer of the poor, John Wright, presented several orders for the cheapest cloth available to be sent to the workhouse and its inmates.[102] In addition to those two very disparate customers, this firm sold goods to a cross-section of the population; church warden, barber, reed-maker, bookkeeper, shopkeeper, 'Irish' woman, coalman, whitster, weaver, fish-woman, packer, and esquire alike, bought textiles and textile goods of every description, ready-made and by the yard.[103] In many cases the orders were modest, as was that of George Piedford, weaver, who bought a silk-and-cotton gown for sixteen shillings. The sum of money was not insubstantial in itself, only in comparison to the orders for hundreds of pounds which this firm dealt in weekly. On occasion some of the seemingly humble customers made surprisingly substantial purchases. A 'Fishwoman', for example, bought nine ready-made gowns one spring and she paid cash.[104]

The entries in the Day Book reveal as wide a variation among trade customers as among the private ones. It is possible to separate trade from private shoppers by the frequency and size of the orders, although there appears to have been no difference in the prices charged. Elizabeth Atkinson appeared frequently in the ledger, and the small size and regularity of her purchases leads to the conclusion that she ran a small retail trade of some kind. In fact, the 1773 *Manchester Directory* identified Atkinson as a milliner and linen-draper with a shop on Smithy-door Street.[105] The orders she presented to the firm were placed routinely. One November in 1775, for example, Elizabeth Atkinson bought one silk-and-cotton gown for seventeen shillings and

---

[102] Ibid. 226, 291, 299.
[103] Ibid. 265, 293, 312, 320, 324, 436, 639, 641.
[104] Ibid. 436.   [105] *Manchester Directory . . . 1773*, 6.

the next week ordered two more. The pattern remained the same throughout all her dealings with the Manchester firm. Seemingly Atkinson did not carry much stock, buying as and when goods were needed. Over a period of one year, from the autumn of 1775 to the autumn of the following year, Elizabeth Atkinson bought only £32. 6s. 10¾d.-worth of merchandise. Most of the money spent went towards the purchase of ready-made gowns, though in addition various types of handkerchiefs and several sorts of textiles were bought to supplement her stock.[106] In this, as in many other cases, the Manchester firm acted as both manufacturer and wholesaler. Undoubtedly Elizabeth Atkinson received a saving from the elimination of delivery charges and the profits of other middlemen. But this favourable practice was made possible because of her proximity to the source of supply in Manchester.

The Day Book of this company gives a further indication of the geographical reach of an individual Lancashire firm. The proprietors dispatched point-to-point shipments to over sixty known destinations in Great Britain and Ireland. (See Appendix 4 for listings of the customers.) In addition the firm sold goods to merchants outside the British Isles in both Amsterdam and Philadelphia. From the many names inscribed in the ledger it has been possible to identify and place almost 170 customers. This catalogue of names does not comprise the total served, only those about whom information was given, or whose stature in the fields of commerce and manufacture ensured some type of public record. Whether or not this firm employed travellers, or whether the owners themselves travelled in search of orders cannot be determined from the existing records. The practice of others would suggest that agents, representatives, or the owners themselves contributed to the breadth of the enterprise.

Trade within the cotton industry featured prominently among the ledger sales in this firm. Nearly fifty of the people and firms identified geographically resided in Manchester itself. Cottons of every available width, texture, consistency, colour, and pattern were exchanged among the dealers as each attempted to acquire the widest selection of merchandise. Each tradesman, manufacturer, or warehouseman had his own set of clients and an order-book to fill. To retain the patronage of their customers they had of necessity to look to other producers for the range of fabrics not of their own manufacture. Robert and

[106] Day Book, 281, 304, 322, 366, 386, 390.

Nathaniel Hyde, for example, manufactured all sorts of cotton, linen, and cotton-linen checks. However, in the diverse consignments from the Hydes to James Beekman, their valued customer, there were in addition to checks 'Fine Dark Cotton Gowns', 'Olive Velverets', women's 'black Mitts', 'Colour'd Thread', and 'Shoe Bindings', among other things.[107] The Hydes did not produce all of these items. But to satisfy customers and keep business healthy they would have scouted round the warehouses of other manufacturers, placed orders with them, and ensured that the required goods were found to fill their clients' needs. The Day Book of the Manchester firm records many small but regular purchases by Robert and Nathaniel Hyde over the years.[108] Manufacturers and tradesmen familiar from other records were also to be found in the ledger of the Manchester firm. J. & N. Philips & Company bought six silk-and-cotton gowns in August 1773; while Abraham Dent's suppliers, John Lawrence and Messrs Barton, traded for years with this firm, during which time they may even have bought goods which eventually found their way to Kirkby Stephen. Manchester check manufacturer Edward Place bought fabric from the firm, as did fustian manufacturer John Hardman and check-maker William Hanson.[109] Just as a retailer might buy from another retailer, so too might textiles travel between two or more wholesalers before reaching the retail distributor, as individual tradesmen added to their inventory. These transactions ensured the standardization of supplies held by wholesalers and secured the homogeneity of available goods to customers throughout the country. Textile tradesmen stocked more kinds of goods in response to demand and looked to other producers for articles which they themselves did not make. Buying, selling, and trading among the chief manufacturers and warehousemen created a pool of common commodities to feed the nation-wide fashion-conscious market.

The firm of fustian manufacturer and chapman Samuel White & Company was one of the most regular customers in the day-book, with orders appearing right through the period encompassed by the ledger.[110] An examination of the purchases made by this company reveals the breadth of products deemed necessary for a perambulatory trade. As well as mundane items such as the linen and cotton checks,

[107] Invoices, 12 Mar. 1767, 10 Jan. 1768, 20 Mar. 1771. Beekman Papers.
[108] Day Book, 53, 410, 423, 425, 426.
[109] Ibid. 1, 2, 19, 312, 343, 368, 382, 390, 472, 474, 487, 530, 537, 539.
[110] White's business is described in the 1773 *Manchester Directory*, 51.

check handkerchiefs, and ticking, White bought crimson and scarlet plaid cottons, worsted plaids, silk damask, velveret, fashionable mallabar handkerchiefs, and inexpensive ready-made gowns, followed in several months by more-costly moree gowns.[111] Samuel White spent a total of £655. 14s. 1¼d. with the Manchester firm in the thirteen orders presented during 1776. The amount and variety of goods ordered by a known chapman confirms the sorts of goods available to rural Britons. White would bring his own utilitarian fustians to rural customers, but in addition there was an unlimited assortment from the other manufacturers. The active sale and trade of products within the community of tradesmen ensured a homogeneous distribution nationally, reinforcing tastes and standardizing demand. M. J. Freeman has speculated that, 'the use of cotton clothing may possibly be dated as far back as 1780 but there were likely marked regional variations in its spread'.[112] Study of the domestic market reveals that the date for general use and wear of cottons can be pushed back at least ten, if not twenty years beyond this. Moreover, the public adoption of cotton textiles and clothing came about in part because of the effective distribution-system sustained by a domestic demand, which was national rather than regional in scope.

The unknown Manchester firm dispatched consignments to sixty-three destinations in the British Isles, in addition to shipments sent to London and trade within Manchester itself. Customers were situated in most areas of the north, including Scotland, central and western Britain, the Isle of Man, as well as in the West Country and Ireland. At first glance the radius of distribution might appear scattered and unconnected. A second look shows the dispersal of orders along the clear and improving transportation-grid of turnpikes and major routes that interlaced the country. Not surprisingly, the pattern of ordering varied among the customers listed in the Day Book. The London and local Manchester merchants operated on a different set of market criterion, where strong demand and easy access to the stock resulted in a more-or-less constant series of orders. For others, the frequency with which orders were placed tended to be a function of distance as well as of the size of the market served. Taylor & Almond were Nottingham tradesmen who provide an apt illustration of this phenomenon. As prosperous provincial tradesmen they made regular

[111] Day Book, 269, 305, 311, 313, 329, 342, 352, 360, 369, 380, 386, 399, 493.
[112] M. J. Freeman, *A Perspective on the Geography of English Internal Trade during the Industrial Revolution* (1982), 14.

and routine purchases, placing fourteen orders between November 1776 and November 1777. These consignments were received at intervals of from two or three weeks throughout the year, establishing an intermediate pattern of ordering, confirming both the ease of delivery and the steady demand of their regional market.[113] Distance had a greater impact on the ordering-patterns of Glasgow customers William and John Dowglass. The first mention of this Glasgow customer appears near the close of the Day Book, but even in the limited time in which their purchases are listed it is possible to observe common features with other long-distance customers. Routine carrier-journeys to Scotland were scheduled several times weekly by this time.[114] However, the lengthy trek involved more time and, in the winter, greater hazards than trips along more populous and heavily travelled routes. During the summer months Scottish customers would receive their orders within one or two weeks, while deliveries during the winter were at the mercy of the weather. Commonly, orders from distant customers were larger and usually placed at quarterly or bimonthly intervals. Examination of the business conducted by Benjamin Bothomley illustrates the correlation between distance and the pattern of ordering. In this case the customer was situated in Amsterdam. Benjamin Bothomley, and later his widow, traded with the Manchester firm over the whole period of time covered in the Day Book. In the twelve months from November 1775 only six orders were placed. Four fell in the months of winter and spring, while one was in summer and the other in the autumn of 1776. Bothomley spent just over £595 on the various textiles.[115] For the customers at some distance from the source, the frequency and size of the orders had to take into account delivery-time as well as possible hazards that could delay consignments and interupt or inconvenience business. The size of William and James Dowglass's orders were very similar to those of Bothomley, though the timing of the Dowglass orders differed. In August, October, and November 1777 the partners received three orders from Manchester. The first, sent by carriage, was for £112. 9s. 10¼d. of cotton check; the second, for £75. 2s. 1¼d., paid for a quantity of mixed quality cottons; while the last, £343. 16s. 4¼d., covered the cost of a huge shipment of mixed cottons of high and low

---

[113] Day Book, 402, 410, 416, 436, 438, 451, 457, 467, 477, 481, 488, 508. Taylor & Almond were also customers of Samuel Oldknow one of the earliest cotton manufacturers. Unwin, 246. [114] *Universal British Directory*, iii, 782.
[115] Day Book, 280, 290, 303, 337, 354, 385.

quality.[116] Glasgow tradesmen could not afford a mischance that could leave their stock depleted; thus the orders tended to be bigger than those of all-but the largest London dealers.

The London route was one of the busiest, offering numerous carrier-services with many optional schedules. On the road south there were also set stops to off-load parcels and make collections. The route Pickford followed ran from Manchester through Pynton, Macclesfield, Stony Stratford, Dunstable, and on to London. The small package of cloth sent to Woburn, Bedfordshire, for example, would have passed along most of the London route, leaving the wagon between Stony Stratford and Dunstable to be transferred to a local carrier.[117] In 1777 Pickford had three trips weekly leaving Manchester and the metropolis simultaneously; by 1788 wagons departed every day but Sunday, and the time the journey took had been halved.[118] A glance at the map of the area served by this Manchester firm reveals a wide expanse of eastern and southern England devoid of its sales. London is notable as being the solitary destination in that wide triangular area of the south-east. A steady flow of Manchester textiles passed along the road to London, but almost no other destination appears within nearly one hundred miles of London—the exception being the single delivery to Woburn and one other to Kent. The commercial structure developed by this firm focused virtually all mercantile activity in that corner of England through the capital. This structure was not characteristic of all Manchester enterprises, as we have seen above with Samuel Ridings. However, the economic shadow cast by the metropolis did tend to draw merchant and commercial men to London for the unsurpassed choice of goods and services.[119] The effects of London's commercial hegemony were felt most powerfully in the neighbouring counties and even in those areas in indirect proximity to the metropolis. These ties were strengthened further by the amelioration of road access to and from London. Thus, the mounds of textiles, hose, and other cottons shipped to London were consumed by a wide regional market served by the retail and wholesale capacity of the city. London newspapers abounded with notices of auctions of textiles and

---

[116] Day Book, 482, 507, 514.   [117] Ibid. 326.
[118] Turnbull, *Traffic and Transport*, 21–3.
[119] Lowndes's 1797 *Directory* notes among the many tradesmen in London 'Boot & Shoe' warehouses, slopsellers, Coventry warehouses, Stafford warehouses, Birmingham warehouses, Hardware warehouses, Nottingham warehouses, and even a 'Chip & Leghorn Hat' warehouse.

other wares aimed to attract tradesman from surrounding counties. Remarks in advertisements such as 'worth the attention of country mercers and merchants trading abroad',[120] accompanied many itemized lists of textiles for sale placed by drapers, warehousemen, and others in the trade. The distributive profile of this unknown Manchester firm may not have been characteristic of all Manchester wholesalers, but the absence of regular delivery-points in the vicinity of London suggests the powerful commercial force exerted by the largest city in Europe.

The London-based Manchester warehousemen figured prominently in the chain of distribution and had distinct features among those entered in the day-book. Many of those whose names were listed repeatedly in the ledger were found to be located in the area of the City around Cheapside, already identified with the trade in cotton products. The following are the names, titles, and addresses of the principal London customers of this Manchester company:

Daniel Cookson, Manchester warehouseman, 39 Lothbury Street;

H. H. Deacon, Manchester warehouseman, 14 Milk Street;

John & Edward Kenworthy, warehouseman, 14 Ironmonger Lane;

Kettle & Mandeville, Manchester warehousemen, 23 King Street;

James Mangnall, Manchester warehouseman, 76 Aldermanbury;

Marsh, Reeve & Co., Manchester warehousemen, 33 Cateaton Street;

John Nickson, Manchester warehouseman, 29 Ironmonger Lane & 4 King Street;

John Shuttleworth & Co., Manchester & Wigan warehouseman, 33 Lawrence Lane;

John Augustus Streit, Manchester warehouse, 346 Strand;

Ellis Needham, warehouseman, 29 Milk Street;

Edward Rogers, Manchester Warehouseman, 78 Fleet Street;

Yates & Miller, Manchester warehousemen, 8 Milk Street;

William Robinson, linen-draper, 74 Holborn;

Lewis & Worsley, linen-draper, 139 Cheapside.[121]

Several of the merchants listed above engaged in some of the largest trade noted in the Day Book. Moreover, of the twenty-four Manchester warehousemen in London in 1776 the firm traded with eleven. The vast metropolitan area with its associated regional markets drew wagon-loads of goods, fabrics, and ready-made items, from the

[120] Ibid. 4 Oct. 1779.
[121] Lowndes, *Directory* (1776), 37, 45, 96, 109, 119, 120, 139, 147, 148, 157, 184.

Lancashire warehouses to those in the City of London. Kettle & Mandeville, J. & E. Kenworthy, and Marsh, Reeve & Company showed the greatest frequency and volume of orders. Kettle & Mandeville had forty-seven orders filled between September 1775 and September 1776, for a total expenditure of £1,234. 14s. 3d. During the year running from September 1777 to 1778 another forty-five shipments were sent south to London for a total of £1,279 7s. 7½d. In the six months from September to March of 1777–8 Marsh, Reeve & Company ordered eighteen consignments of textiles at a cost of £320. 2s. 6½d., which, while not comparable to the level of orders sent by Kettle & Mandeville, still reflected a vigorous trade.[122] In the case of Kettle & Mandeville the ledger reveals that one order per week could be expected during their busiest season. The volume of this trade is a far cry from the provincial commerce of Abraham Dent, who in 1767 spent a total of £677 on all his merchandise.[123] Set against this modest concern, the scope of the London trade becomes apparent.

One of the most popular items among the London shop-owners were the gowns which came in several fabrics and a multitude of prints and colours. Sample-cards probably presented the fabrics, and perhaps a sketch of the gown as well, for their perusal. Some would sell only dress lengths, but shopkeepers in urban settings seemed to have an inexhaustible desire for ready-made gowns. The sale of ready-made gowns by both large and small retail shops in London was well established by this time, and the rate at which orders flowed into the metropolis indicates the turnover of stock in their establishments.[124] In April 1776, Kettle & Mandeville bought fifty-one gowns and ordered seventeen more in the following month; while J. & E. Kenworthy had delivered sixty-nine gowns of varying types over the same period.[125] Throughout the year from September 1777 to September 1778 Kettle & Mandeville bought over 800 ready-made gowns of different fabrics, colours, and prints, all of which were shipped to their London warehouse.[126] However, these products were not always required at an even rate. On examining the entries for Kettle & Mandeville it is apparent that the shipments did not arrive at a uniform and unvarying pace. In the distribution of textile products to London, as in the movement of merchandise elsewhere in the country, patterns emerged that were peculiar to the region being served. A recurring cycle of

[122] Day Book, 261–375, 492–581.   [123] Willan, *Abraham Dent*, 28.
[124] The trade in ready-made clothing is discussed in more detail in Chapter 5.
[125] Day Book, 321–2.   [126] Ibid. 492–581.

orders characterized the business practice of Kettle & Mandeville, with expenditures rising markedly in the early autumn, remaining high in December through to the early spring, to drop sharply in the summer months. This pattern appeared during all the years catalogued. Highs and low such as these are characteristic even today for certain products. At this time the cyclic lull in shipments from Manchester to London mirrored the seasonal fluctuations in the surrounding countryside, when the focus of attention was shifted in the spring from the affairs of the capital to those of the countryside. The gradually escalating rate of orders through the autumn and winter corresponded to the reawakening of the city after the summer hiatus; the resurgence of political, legal, mercantile, and social rounds that accelerated with hardly a break into the spring.

London's Season involved not only the immediate participants, but the thousands of subsidiary players and suppliers, underpinning the revolving social pantomime. The commercial stimulus resulting from these elaborate entertainments spread through every layer of the city. This boost was compounded by the influx of seasonal visitors intent on Parliament, law-courts, business, and trade, as well as pleasure. A general rise in consumer spending at every level resulted. Increased demand would then be translated into larger and more frequent orders from the London warehousemen to their suppliers. Robert Owen described the rigours of his apprenticeship to the drapers Flint & Palmer, of London Bridge, the most noteworthy being the seasonal fluctuations in trade during the peak spring period.

Between eight and nine the shop began to fill with purchasers and their numbers increased until it was crowded to excess ... and this continued until late in the evening; usually until ten, or half-past ten, during the spring months ... When the spring trade ceased, and the business became less onerous, we could take our meals with some comfort, and retire to rest between eleven and twelve, and by comparison this became an easy life.[127]

London districts were touched to a greater or lesser degree by the fluctuations in trade, but all felt the effect to some extent. So great was the London market that the ripples were felt too in the manufacturing districts of Lancashire.

Although the commonest mode of transporting bales and boxes of textiles was by carrier, occasionally orders were also forwarded by coach. Only 2.6 per cent of the 1,035 orders dispatched in 1776 went

[127] Robert Owen, quoted in Alison Adburgham, *Shopping in Style* (1979), 43–4.

by coach, but of that total two-thirds were bound for London. The benefit of sending limited consignments by coach was obviously the speed with which they were delivered. Travel-time had certainly decreased for the wagons lumbering down to London. But the dashing coaches cut down the time even more radically than did the sluggish carriers. The 'Flying-Machine' sped along the route in less than a day-and-a-half in 1780, a trip that had taken three days in 1760.[128] In 1793 the journey to London could be completed on the Royal Mail coach, weather permitting, in a breathtaking twenty-eight hours.[129] The diminution of the journey-time tied the production centre more closely to the London market, helping manufacturers become more attuned to the demands and requirements of their clients. The ledger entries of 1776 reveal that in all cases the cargoes had a value sufficiently high to ensure a return to the retailers in spite of the extra cost of coach shipments. Here again the number of times the firm resorted to coaches corresponded to the acceleration of ordering during the winter months. Indeed, the orders shipped by coach during the months of November and December outnumbered the total number of coach-bound shipments in all other months.

Ready-made gowns ranked high in the list of items urgently requested. (See Plate 19, where Gedge, linen-draper, announces his Manchester products.) Undoubtedly Kettle & Mandeville reflected in their ordering the requirements of their retail customers, who made specific requests for the red moree gowns which appeared in the order-book the month before Christmas 1776. The red gown had an immediate success and Kettle & Mandeville must have hoped to build on this new fad. Speed was of prime importance in the provision of this holiday attire, for the sooner the clothes arrived the sooner they could be displayed in the shops and profits rung in with the season. Kettle & Mandeville were sent the first consignment of red gowns on 17 December; further shipments were forwarded in all haste on 24 and 31 December. Rapid delivery was essential to capitalize on the transient seasonal demand for a gown of this colour, which cost one shilling more than the regular twenty-two shillings for a plain moree gown. So while a large load of textiles and assorted gowns began the journey to London by wagon, the parcels of specially dyed moree gowns sped on ahead.[130] Samuel Salte's letter to Oldknow encapsulates

[128] W. Harrison, 'The Development of the Turnpike System in Lanashire and Cheshire', *The Lancashire and Cheshire Antiquarian Society* (1886), 87.
[129] *The Universal British Directory*, iii, 780.   [130] Day Book, 405, 409, 411.

the urgency he felt at a dearth of special cotton fabrics; an urgency that might have been shared by many merchants at one time or another:

press forward in all the finer Articles & as fast as possible... We want as many Spotted Muslins and fancy Muslins as you can make the finer the better.... Send by the Coach everyday what you can. This Month & the next are of the uttmost [sic] consequence to you & to us. We expect to hear from you as often as possible & as the Sun Shines let us make the Hay.[131]

Speed and profit were often synonymous in the London market. So eager were some manufacturers to promote their products in the London market that they offered to deliver them free of charge.[132]

The Manchester firm sent few consignments by coach aside from those to London. Of the provincial orders routed by coach, a handful were high-quality textiles or ready-made gowns. Cargo sent by coach was in general composed of medium-quality mercery goods: 'Cotton Holland', 'Cotton Checks', 'Cotton Plod [plaid]', and 'Sheetg dyed blue'.[133] The best-quality fabrics were the 'Plain Silk Velvets' and the 'Feathered Tabbynetts' sent by coach to Richard Davies in Exeter.[134] The order sent from Manchester to John Thomas of Hinckley suggest the different use of coach transport outside the London region. Only one of the invoices notes an order sent by coach. A manufacturer had been unable to fill Thomas's earlier order, and on receipt of the materials sent the two outstanding pieces of fine cloth to Thomas by coach, with the larger order to follow by carrier.[135] Clearly, use of coach transport outside the Manchester to London corridor was approved only in very special circumstances, whereas the peculiar market-forces extant in London necessitated a more flexible response. In all but a few cases provincial customers around the country were content to rely on the somewhat slower deliveries by wagon and cart.

The cost of transportation by carrier was an important component of the total price of the cotton goods distributed throughout Britain. Both the retail price of cottons and the profits for the distributor were influenced by the cost of shipping cloth to the retailer. Eric Pawson summarized the analysis of Jackman and Albert on the general movement of carriage rates, based on the limited documentation available. The assessments of the three historians concur. Between 1750 and 1800 the cost in real terms of road-haulage dropped,

---

[131] Salte quoted in Unwin, 64. [132] Unwin, 64
[133] Ibid. 310, 482. [134] Ibid. 607.
[135] Eng. MSS 1192, Invoice, 4 Oct. 1786, John Rylands Library.

indicating that the faster trips which resulted from improvements in road surfaces and carriers' equipment offset the rise in feed-costs and industrial prices.[136] A broad consensus such as this is important in clarifying the economic environment in which the manufacturers and tradesmen functioned. More specific indexes of the cost of transportation and the percentage of this cost in the retail price of cotton goods would be of even greater value. Unfortunately, the evidence is not available that would make that sort of appraisal possible. The ledger entries for the anonymous Manchester firm indicate shipping costs only when goods are not sent by wagon. These orders, being exceptional, warranted an additional notation; whenever the method of shipping varied from the norm it was noted.[137] But for the vast majority of orders the shipments were sent on the routine carrier-runs. The Day Book did not record the standard rates charged, or even the individual carriers employed year in and year out. Without the exact costs of the shipments to the various destinations, and the cost:weight ratio, it is impossible to apportion the percentage cost of transportation to the final price of the goods.

Only one series of entries in the Day Book provides any additional insights into the question of transportation. James Gee was a customer who made large and regular orders with the Manchester firm. One such typical order was for fabrics of various sorts to the value of £189. 9s. 3¾d. At the end of the entry the clerk indicated that Gee had received a credit of fifteen shillings on the order because he had collected it himself, rather than having it delivered.[138] James Gee placed further orders with the Manchester firm, and whenever the merchandise was collected the total was reduced by an amount that appears commensurate with the transport costs to Stockport, where Gee was a resident. Gee was the only customer to collect his large orders, which were composed exclusively of bulky bolts of cloth. In so doing he appears to have eliminated the costs of a commercial carrier, substituting instead his own transportation. These incidents demonstrate the cost of moving high-bulk cargoes over a comparatively short distance. However, without a corresponding chart relating the various

---

[136] Eric Pawson, *Transport and Economy* (1977), 296–7.

[137] One example of a method of transport unique in the ledger was the use of coastal vessel to transfer a cargo from London to Plymouth. The Plymouth merchant had ordered a very cheap, coarse linen-cotton cloth and the lowest-cost transport was indicated. Under the entry was written 'by Cooper [a London carrier] to Chamberlains Wharf London from thence by first coaster, Carrg pd 3/10d.' The order cost £3. 16s. 1½d.

[138] Day Book, 335.

pieces of cloth to weight and weight to cost over distance, once again the information is only suggestive of the sorts of expense involved in the movement of consignments around Britain. Moreover, where key sales could be made to London wholesalers, manufacturers were ready and willing to absorb the cost of transporting goods to market.[139] The expense in this instance came second to possible mass sales.

The textiles manufacturers of Lancashire and the surrounding counties depended upon the network of turnpike and post-roads binding the country together. Once passable roads were assured, the manufacturer did not have to depend on the ponderous seasonal tours of the provinces to facilitate sales. Neither did the local retailer have to accept whatever was brought him, with the view that any stock was better than none at all. Extensive trading among manufacturers, wholesalers, and retailers, the use of travelling representatives and manufacturers samples, forged closer ties from manufacturer to consumer—demand was more easily telegraphed along this progression of interests. The collective sales of the pedlars, hucksters, and travelling Scotchmen persisted as the only effective means of access to much of the remote rural population. As villages grew and cities swelled, however, it was the requirements of the settled retail vendors that stimulated new methods of soliciting sales. Manufacturers quickly learned from the orders sent by post or placed with their travellers which of their products the public favoured, adjusting production accordingly. Moreover, with consumer goods so quickly delivered to town and village the retail trade received a tremendous boost. Shopkeepers and tradesmen, no matter how distant from the production site, had available to them a wide and growing supply of manufacturers from the British cotton industry.

However vast the warehouses and numerous the merchants in the capital, London did not succeed in monopolizing the domestic sale of British-made cottons in the way it had the Indian precursors. Manchester reigned in tandem with London as distribution-hubs for the cotton industry. Furthermore, none of the provincial manufacturing towns competed with this regional specialization. Towns such as

---

[139] Salte wrote to Oldknow in April 1786 of several manufacturers who had begun shipping goods free of transport charges. 'Strutt sells his at 20½ Free of Carriage, another Gentn at Stockport offered us this [and] any Quantity of fine White Callicos at 20d. free of Carriage.' Unwin, 64. This offer can probably be explained by the crisis of production and the growing competition in 1786. But it is a suggestive practice, perhaps indicative of the declining costs of transporting textiles along well-serviced routes.

Blackburn, some distance from Manchester, with about sixty cotton manufacturers resident in the town, might be expected to attract some warehousemen: in fact, not one warehouseman appeared in the town directory at the end of the eighteenth century.[140] The selective clustering of pedlar firms around certain Lancashire towns did not challenge, but rather supported the developments in this trade. The concentration of travelling hucksters in many cotton towns was almost the sole distributive activity to thrive outside the London–Manchester axis. By the end of the century, only two other urban sites in England were discovered to have traders identified as Manchester warehousemen. These two locations were at right angles to the powerful London–Manchester commercial axis. One Manchester warehouseman remained in Bristol at the end of the eighteenth century. The northeast offered more fertile ground for local merchants, and three Manchester warehousemen were resident in Newcastle in the last decades of the century.[141] Earlier in the century both Bristol and Newcastle attracted warehousing facilities for the regional distribution of Manchester goods; decades later this supplementary capacity still remained. These were the exceptions. The absence of other contending, or even complementary centres of distribution confirms the tremendous commercial authority concentrated in Manchester and London. The opposing geographical positions of these two sites assisted in perpetuating the hegemony of the two very different cities, as the main dispersal-points for the British cotton industry. Throughout the country, shopkeepers served by the several means of distribution were able to offer their customers a homogeneous selection of textiles consistent in content from one side of the country to the other.

[140] *The Universal British Directory*, v, 18.
[141] *Bailey's Northern Directory* (1784), 249–53.

5
MARKETING FASHION: COTTON CLOTHING AND THE READY-MADE CLOTHES TRADE

BY YOUR DRESS YOU SHALL BE KNOWN?

SOCIAL identification through clothing was for centuries one of the underpinnings of European society. It was assumed that a quick scanning of the stranger would provide sufficient clues to status, occupation, and perhaps regional association so that few errors would be made during social interaction. Dress continued to be seen as a safe and accurate gauge of rank long after the formal prohibitions against dressing above one's station were swept from the statute book in England, early in the seventeenth century. However, expectation and experience did not always coincide—a problem for moralists and all those who sought the reassurance of a permanent cultural norm. The fashion-sponsored changes in attire were assumed to be the exclusive prerogative of the gentle classes, even though legal sanctions had eroded. Commerce, however, made fashionable dress less a signal of the courtier's standing or the landowner's station, and more a commodity in the public domain, access to which was determined through objective criteria. By the early modern period in Britain outward show became marketable. The tide of commerce engulfed the margins of a once-exclusive domain and swept the rights to fashion from this select circle to the wider world. Dress was for sale, and fashionable dress was for sale also.

A simple statement such as this masks profound alterations in the fundamental assumptions held by members of that society. Against the objections of the clergy in their pulpits and the landowners in Parliament, the majority of the consuming population expressed their confidence in a system of commercial practice that made fashionable goods like cottons accessible and progressively less expensive. 'By your dress you shall be known', the moralists, traditionalists, and conservatives might have intoned, with every hope of being able rigidly to apply this dictum. Such hopes were all-too frequently confounded, for in Britain commerce had removed one of the constituents of the aristocratic milieu and transported it, altered and cheapened but with its essence intact, into the common market. Public denunciations of those who transgressed the boundaries of dress represented a rearguard action in defence of a privilege besieged and then overrun. Protests and appeals to morality and tradition did not convince the many that they should foreswear the masquerade of fashion that blurred the divisions of rank. In the prelude to the industrial age a rehearsal of political and social adjustment took place in the realm of dress, as the restrictions on appearance were challenged and cues to rank obfuscated down through the middle and prosperous labouring classes. The lure of cheap attire that was a facsimile of élite fashions, the allegiance to the commercial impulse, ultimately sparked further innovation and industrialization. In the second half of the eighteenth century, production advances made even the question of price less of a stricture on the urge to consume.

In the political arena this struggle between the new commercial and the old landed interests has received great attention.[1] Popularized fashion is a pivotal component of this process worthy of a more intensive inquiry, for in decrying the general wear of certain clothes critics were not simply debating a matter of ephemera. Neil McKendrick was one of the earliest advocates of a reconsideration of the significance of fashions as a reflection of fundamental social change. 'In my view', he wrote, 'the Western European fashion pattern (and indeed the more general Western European consumer pattern) is as marked, as important and as worthy of attention as the much studied "European Marriage Pattern".'[2] Fernand Braudel contended that, 'Costume is language'. As such, it 'is no more misleading than the graphs drawn by demographers and price historians. In fact the future belonged to societies which were trifling enough, but also rich and inventive enough to bother about changing colours, materials and style of costume.'[3] Over a century earlier a fellow-countrywoman, Flora Tristan, arrived at the same conclusion.

---

[1] One example of which is the work done by John Brewer, *Party Ideology and Popular Politics at the Accession of George III* (1976); more recently Professor Brewer has contributed to the literature on the commercialization of British society with 'Commercialization and Politics', in Neil McKendrick *et al.*, *The Birth of a Consumer Society* (1982).

[2] McKendrick *et al.*, 41.

[3] Fernand Braudel, *Capitalism and Materials Life, 1400–1800* (1967), 235–6.

I am firmly convinced that it is not necessary to understand the language of a country in order to understand its ways. They are revealed by every outward sign, and particularly by dress.

As opinions, manners, customs, and fashions are translated into objects and actions, and as they all have their causes from which they stem according to the laws of nature, I maintain there is nothing pertaining to a nation which cannot be understood by alert and thoughtful observation, without the help of the written or spoken language.[4]

Historians are not obliged to forswear the written word in their enquiries. But too often the commonplace, matter-of-fact business of dress has been ignored in favour of more weighty matters. Social differentiation, popular attitudes, and economic activity, are signalled in the ebb and flow of style. One element of this developing materialism is described by J. H. Plumb as, 'the will, the desires, the ambitions, and the cravings of the men and women who wanted change and promoted it'.[5] Within the transitional era in Britain, before the unleashing of great industrial might, one has to attempt to assess the metamorphosis which saw the translation of fashion into a common vernacular. No other realm of economic and social life more cogently reveals the ambitions, aspirations, and appetites of its members than does a study of the popularization of fashion.

Fashion was more than 'the creation of producers'; nor was industrial mass-production the spark that inflamed middle-class interest in material goods, as has been suggested.[6] Outside of the realm of technology, the control of the manufacturer over fashion was at a cosmetic level, providing a stream of alternative products *in response* to a structure of demand that was indivisible from social and political attitudes. The structure and function of popular fashion and generalized demand was rooted in an outlook which recognized the existing hierarchy, but at the same time aspired to the higher ranks within view.[7] The genesis of the cotton industry in Britain was founded

[4] *Flora Tristan's London Journal, 1840,* trans. Dennis Palmer and Giselle Pincetl (1980), 247.   [5] J. H. Plumb, *Georgian Delights* (1980), 10.

[6] Toshio Kusamitsu, ' "Novelty—Give us Novelty": London Agents and Northern Manufacturers', Paper delivered at the Pasold Conference on the Economic and Social History of Dress, London, 1985, 3.

[7] This ambitious behaviour was undeniably an irritant to those mimicked. John Byng, later Viscount Torrington, betrayed in his diary the vexation of the great at seeing the more lowly following close at heel. 'In Hagley village are some neat houses', wrote the diariest, 'all copying the greater example; for wherever a man, or garden of taste is established there are allways around them some imitative warts.' C. B. Andrews (ed.), *The Torrington Diaries* (1886), 47.

on these far-from ephemeral wants. Fashionable clothing was the point of embarkation in this social masquerade. To appear in an acceptably fashionable attire might entail no more than the wearing of a petticoat of the prescribed colour, having a muslin apron suitably embroidered set at the right length, sporting a printed calico handkerchief agreeably draped, maintaining white cotton stockings and linens correctly laundered, or re-trimming a gown to the correct style.[8] These touches denoted the devotees of fashion in whatever rank of society they arose.

Comments both before and during the eighteenth century suggest that the lure of fashionability infected British society.[9] But the frequency and unanimity of these remarks reaches its height in the mid- to late eighteenth century.[10] The visiting Swedish botanist, Per Kalm, considered England a paradise for farmers' women, not only because of the leisure they enjoyed but also on account of the modish dress in which they indulged on Sunday. He wrote of the Hertfordshire farm-wives that,

All go laced, and use for everyday a sort of *Manteau*, made commonly of brownish Camlot. The same head-dress as in London. Here it is not unusual to see a farmer or other small personage's wife clad on Sunday like a lady of 'quality' at other places in the world and her every-day attire in proportion.[11]

[8] 'Tale of a Puce Dress', *Costume*, 6 (1972), 100, provides an example of a dress altered almost yearly from 1781 to 1788; additions of steel buckles, a new gauze handkerchief, and a new apron all extended the life of the garment and retained a stylish appearance. Note also the case of the stolen silk gown that came before a magistrate in 1784. The owner recognized the gown on the thief's back as she approached her on the street. The silk gown was so easily identified because it had been altered many times over ten years. *Proceedings of the King's Commission of the Pence . . . for the City of London and . . . the County of Middlesex; held at the Old Bailey* Oct. 1784, 1,281.

[9] See R. Reuss, *Londres et L'Angleterre en 1700*, describing with wonderment the attire of the urban artisans: 'Their dress is more than luxurious and one sees the wives of tailors and shoemakers wearing clothes embroidered in gold or silver and adorned with gold watches.' Daniel Defoe claimed that 'the working manufacturing people of England . . . make better wages of their work and spend more of the money upon their backs and bellies, than in any other country', *The Complete English Tradesman* (1726), 250.

[10] Neil McKendrick records a host of such comments in 'The Commercialization of Fashion' in *Birth of a Consumer Society*. J. M. von Archenholz wrote in 1787, for example, that, 'The appearance of the female domestics will perhaps astonish a foreign visitor more than anything in London' (p. 57); while the Russian visitor, Nikolai Mikhailovitch Karamazin writing at about the same time was struck by 'a general appearance of sufficiency . . . lord and artisan almost indistinguishable in their immaculate dress' (p. 80).

[11] Per Kalm, *Account of his Visit to England . . . in 1748*, trans. Joseph Lucas (1982), 326.

Kalm considered that English people of all sorts were most extraordinarily preoccupied with the requirements of fashion:

> I believe there is scarcely a country where one gets to see so many *Peruque* as here. I will not mention that nearly all the principal ladies, and also a part of the commoner folk, wear Peruques, but I only speak of the men, who in short, all wore them. . . . It did not, therefore, strike one as being at all wonderful to see farm servants . . . clodhoppers . . . day-labourers . . . Farmers . . . in a word, all labouring-folk go through their usual every-day duties with Peruques on the head. . . . I asked the reason for the dislike of, and the low estimation in which they here held their own hair. The answer was that it was nothing more than the custom and *mode*.[12]

Thus, fashionable indulgences were not solely the prerogative of the female sex. Consider the attention devoted to the question by James Boswell, newly arrived in London from Scotland. He well knew that frugality would not serve when it came to his appearance and calculated to the penny how much he could afford to spend on the maintenance of his person, in the form of linen, laundry, and clothes. On his arrival in the capital, Boswell set out to acquire as modish a demeanour, as fashionable a circle of friends, as his purse and his personality could sustain. He acknowledged that it would not be easy to 'support the rank of a gentleman' on the allowance of £200 per annum granted him by his father, so he devised a 'Scheme of Living Written at the White Lion Inn, Water Lane, Fleet Street, the Morning After My Arrival in London, 1762'. Every expenditure was enumerated, including how frequently he could have fires and whether or not he could afford to host guests for breakfast. On the matter of his attire Boswell was specific:

> I would have a suit of clean linens every day, which may be 4d. a day. I shall call it for the year £7. I would have my hair dressed every day, or pretty often, which may come to £6. I must have my shoes wiped at least once a day and sometimes oftener. I reckon this for the year £1. To be well dressed is another essential article, as it is open to everybody to observe that. I allow for clothes £50. Stockings and shoes I reckon for the year £10.[13]

Fashion's requirements were reckoned precisely and carefully budgeted in Boswell's diary, a familiar practice to those who aspired to a suitably appearance on limited means. In pursuit of his goal he noted

---

[12] Kalm, 52.
[13] *Boswell's London Journal, 1762–1763* (1950), 335–6.

a week after his arrival, 'I had now got a genteel violet-coloured frock suit'. And his efforts were to some effect, as he noted at the end of December next.

I now received a card of invitation to the rout on Tuesday the 7. This raised my spirits, gave me notions of my consequence, and filled me with grandeur. Fain would I have got rich laced clothes, but I commanded my inclination and got just a plain suit of a pink colour, with a gold button.[14]

Such admirable self-restraint and a nice calculation of the norms acceptable to the social set to which Boswell aspired sent him on the way to fulfilling his goals in the metropolis.

Anne Buck has shown that Britain differed from the rest of Europe in the manner in which their citizens dressed. No one group had a specific, exclusive garb. One's social station or type of work did not impose an immutable costume. And although there was 'gradation and overlay between rank and rank with differences in detail . . . no dress is so different that it shows a completely unrelated, independent style'.[15] Throughout most of the eighteenth century the construction of women's garments changed very little, and the styles that there were appeared commonly in all regions and in all classes. Two sorts of garments predominated: one was essentially a two-piece robe, open at the front, worn over a petticoat; the other, popular at the beginning of the century and again in the 1780s, was a closed gown with a fall of fabric from the bodice, over which an apron was worn. The only other type of gown was known as a bedgown. It wrapped over, closing at the front and was about knee-length, usually the dress of working-women.[16] The styles of the gowns were broadly uniform, quality apparel being distinguished by the cut, fabric, and trim: this alone differentiated the humble from the *haute couture*.

As the basic construction of women's attire was standard throughout the country, it was therefore the minutiae of fashion that determined the disciple of 'la mode'. The distribution of lighter attractive fabrics was proceeding apace. Knowledge was the ultimately restricting force. Popular colours, new ways of trimming gowns, of wearing linens, news of the current vogues being worn in the metropolis, were all elements that would distinguish the dowdy from the chic. Without regular, accurate information on the current trends provincials, the great bulk

---

[14] *Boswell's London Journal, 1762–1763* (1950), 53, 65.
[15] Anne Buck, 'Variation in English Women's Dress in the Eighteenth Century', *Folk Life*, 9 (1971), 5.   [16] Ibid. 18.

of the population, were condemned to be forever trailing behind the current modes, by months or years. The potential was certainly there for every woman of modest means to have at least one stylish gown, and similarly for the man, the jacket, waistcoat, and breeches decreed by fashion; the obstacles were the limitations imposed by geography and the delays in the passage of information.

For much of the early eighteenth century the country-dwellers and those living in small towns and villages depended for their knowledge of the latest modes on shopkeepers, pedlars, and others who journeyed to London, the closest provincial centre, or a fashionable spa. Country tradesmen, tailors, and mantua-makers relied on travellers for the current styles, and to this end dolls were routinely circulated to exhibit a scaled-down version of the fashions. For those country-folk attempting to present an elegant appearance, success hinged on the rate at which the new styles could be incorporated into an existing wardrobe. Distance imposed the greatest restraints, unless there was by chance a fashionable peer in the vicinity,[17] in which case a quick eye and a memory for detail could aid in the revamping of a costume. However, that was a hit-and-miss situation. By mid-century no means had yet been found to satisfy the demand for detailed fashion-news among the majority with no direct connection with high society, but with a burning desire to emulate the dress and manners of the metropolitan élite. Isolation from the urban centres and outdated and insufficient information inhibited the establishment of homogeneous fashions. Fashion-related intelligence came more frequently with the faster pulse of trade, and was further accelerated after the establishment of the principal turnpikes. Once better lines of communication were fixed, the flow of all sorts of news, in the form of journals, newspapers, and magazines, poured out to the waiting public.

FASHION-NEWS IN PRINT

Provincial newspapers were well established by mid-century. Under their influence regional tradesmen were brought into closer contact with central suppliers, contributing to the standardization of the

---

[17] The duke of Hamilton was one such peer, who returned from the capital to his estate with fashionable clothing for his wife and himself. Rosalind K. Marshall, *The Days of Duchess Anne* (1973), 86–8, 92–3. Note also, McKendrick *et al.*, 74, examples listed of the lords and gentry newly returned from London and carrying with them the most current styles, to the delectation or chagrin of church-goers in the provinces.

market and the expectation of the consumers.[18] In these papers, advertisements brought news of goods and also might announce the presence of a London supplier. The *Stamford Mercury* carried such an announcement in March of 1728. This retailer placed several ads in the weekly paper to stir interest and bring customers to his temporary setting 'at the Corner of the George and Angel Inn in St. Mary's Street' during the duration of the Stamford Mid-Lent Fair, where he could offer a metropolitan selection to local residents and visitors.[19] These provincial periodicals likewise provided details of significant fashions, such as the correct dress for court mourning on the death of the king. Those desirous of emulating the practices of the court would then have the necessary information to amend their dress.[20] Details of the appearance of those attending court functions, like the queen's birthday celebrations, were likewise related: some of those attending were described as wearing 'new-fashion'd white Lutestring with Diamond Spots, either of Pink, Purple, or Green . . . The Sleeves to their Gowns were not quite so short as last Year'.[21] Even at this the newspapers did not satisfy all the needs of their readers, and ultimately another segment of the printing-trade addressed this problem.

The Stationers' Company obtained a monopoly for the publication of women's almanacs in 1704, and brought out *The Ladies Diary or The Woman's Almanack*. In 1750, a black-and-white engraving of a stylish gown was included at the front of this volume, in response to the interest in fashions evinced by its readers. Pictures of this sort were included in all subsequent editions, depicting styles of full dress, undress, head-coverings, bonnets, and accessories. In 1770 the Company's monopoly was successfully disputed by other publishers and a torrent of pocket-books and memorandum-books were produced, all with engravings of fashionable figures displayed on the front pages.[22] These small volumes found an immediate audience. They

---

[18] An example of this sort of advertisement can be found in the *Stamford Mercury* for 6 June 1728: 'Barnaby Turner, Hatter, at the Hand and Hat over-against the Black Swan in the High-street, Stamford, being just come from London, hath very good choice of all sorts of Mens and Boys black Hats; also all sorts of Womens and Girls, viz. Black, White, Blue, Cane, Straw, Leghorn, Silk and Velvet Hats for the Ladies, with all sorts of Hunting and Silk Caps for Men, Women and Children. Also fresh Coffee and Teas of all sorts out of the last Sale, Chocolate, Cocoa, Scotch and Spanish Snuff, with Variety of Holland, London and Tunbridge Toys for Children, fresh come down, and all at lowest Prices.'   [19] *The Stamford Mercury*, 21 Mar. 1728.

[20] Ibid. 15 Aug. 1728.   [21] *Northampton Mercury*, 8 Mar. 1735/6.

[22] *Hollar to Heideloff*, catalogue from an exhibition of fashion plates, Costume Society (1979), 21–2.

circulated all around the country, illustrating established modes for those with no other models to follow. The small engraved figures in the women's pocket-books were the first fashion illustrations devised for that purpose in Britain.[23] (See Plate 20, six examples of these plates.) These books became the authoritative guide to dress, however long the delay between the advent of the style and its arrival in the country. Their limitation lay in the months between the illustrator's pen and the appearance of the edition in provincial shops, as well as the fact that only one illustration was produced each year. Not surprisingly, the rustic devotees of the almanac or memorandum-book became the subject of mockery, an example of which is found in Oliver Goldsmith's *She Stoops to Conquer*. In this play the proud provincial, Mrs Hardcastle, makes her claim to fashionability known to the London visitors by announcing that she has arranged her hair in the manner of 'a print in the Ladies' Memorandum-book for the last year'—a style outmoded to the callers and an invitation to laughter at her pretensions.[24] The demand for fashion-news was addressed by the memorandum-book publishers; but the problem of accuracy could not be solved in this format.

The Gentleman's Magazine inspired the publication of several similar magazines for women, most of which were short-lived. One of these journals, *The Lady's Magazine*, edited by Oliver Goldsmith, ran from 1759 to 1763, and in the first year a full-page black-and-white engraving was issued. Entitled 'Habit of a Lady', it was the first fashion-plate to be issued in an English monthly magazine, 'For the assistance of those in the country who, as they have not the opportunities of seeing the originals, may dress by the figure'.[25] The publishers recognized the value of this sort of endeavour, as well as the extent of demand from the literate middling classes for information that would allow them to present a heightened social appearance. In 1770 the first volume of *The Lady's Magazine; or Entertaining Companion for the Fair Sex* appeared in bookshops. Advertisements for this new periodical appeared in newspapers in the provinces.[26] It was a widely read journal, being the only one of its kind in England at this time. The spread of fashion-news began to achieve a momentum of its

---

[23] Ibid. 22. Although in France fashion sketches were published in the late seventeenth century in periodicals like the *Mercure Gallant*. These publications were brought over to England, where they found an eager audience.
[24] Oliver Goldsmith, *She Stoops to Conquer*, Act II, Scene i.
[25] *The Lady's Magazine*, Dec. 1759.
[26] *The Bristol Gazetteer*, 8 Feb. 1776; *Exeter Flying Post*, 4 Jan. 1792.

own. Fewer critics disputed the right of readers to know and apply the information received, although the voice of traditional conservatism was not completely stilled. In publishing circles there was growing competition to see who would furnish the fullest and most complete reports of current styles. The editors of the *Lady's Magazine* had a clear appreciation of their role in the extension of fashion throughout Britain and of the unbounded curiosity of their readers. The first issue of the journal laid out editorial policy with regard to the great moral, social, and economic question of fashion:

But as external appearance is the first inlet to the treasures of the heart; and the advantages of dress, though they cannot communicate beauty, may at least make it more conspicuous, it is intended in this collection to present the sex with the most elegant patterns for the Tambour, Embroidery, or every kind of Needlework; and as the fluctuation of fashion retards their progress into the country, we shall by engravings inform our distant readers with every innovation that is made in the female dress, whether it respects the coverings of the head, or the clothing of the body.[27]

The first issue of *The Lady's Magazine* contained a colour-engraving of 'A Lady in Full Dress', but other fashion-plates appeared only at irregular intervals until 1780, when they were produced each month. Purists in the study of fashion-plates contend that the first decade of the journal showed only a scattering of fashion-engravings,[28] but for our purposes there were a far greater number of illustrations than this statement would suggest. Every edition of the magazine included short stories. Adjacent to these moral tales were engraved plates, most frequently picturing exquisite young men and women, matrons, or children, in dramatic postures. In almost all of these vignettes the models were portrayed wearing the most fashionable clothing consistent with their position. (See Plates 21 and 22, from *The Lady's Magazine* of the 1780s.) From these illustrations readers of both sexes could ascertain what was being worn in London. Items could then be bought to modify existing clothes or new suits of clothes purchased to reflect the modes of the metropolis. The homogeneity of taste and material culture would be extended further by the judicious pirating of engravings by rival publishers. One engraving, first presented in *The Lady's Magazine*, became so popular that, wrote the editors, 'we find it imitated by most of the editors of the annual pocket-books for the use

[27] *The Lady's Magazine*, 1 (1770), introduction.
[28] *Hollar to Heideloff*, 30.

of ladies'.[29] Thus the costume copied by the engraver witnessing various events on the social calendar, would be reproduced in publications around the country and consulted by their patrons as the models for their own dress.[30] Simultaneously, an expanding fund of moderately priced British cottons made all these styles more easily replicable.

The Lady's Magazine added further to the amount of fashion information offered their readers with small monthly dissertations. For February 1773 it was announced that an undress should consist of:

The hair in front, with small puff curls; a cap, made of wings; narrow ribbon, in small puffs; a double row of lace; ditto lapelled, double and puffed; a very small crown of blond, lined, and coming so high behind as to admit a very low bag of hair; gown as rich of sattin or tabby as the lady likes ... short black apron of joining lace; double ruffles very deep; a flounce at the bottom of the gown ... round the neck a very narrow collar ...[31]

A rather less elaborate toilet was suggested for the undress in 1775: 'All sorts of worked gowns over small hoops. ... Night gowns in the French jacket fashion, flying back, and tying behind with large bunches of ribbon. Sashes round the waist, and fastened with a small buckel. Short Aprons. Shoes with buckels'.[32] Thus the readers were advised to adorn themselves.

Interest in the fashions of the day was not restricted exclusively to the women's journals and pocket-books. Significantly, it was in a western rural region very distant from London, in the early years of fashion publication, that there appeared a regular bulletin on the London styles for the information of the prosperous local shopkeepers, farmers, landowners, and other readers. Those living in close proximity to a spa or to other commercial or provincial centres might not have required this type of detailed instruction. Residents in those more cosmopolitan locales could visit modish shops and suppliers, or model their attire on the resident élite; advantages not shared by those immured in a bucolic environment. *The Sherbourne Mercury* produced in its appendix a segment entitled 'The dress of the Month as established at St. James's, and in Tavistock Street'. What followed was a description of the latest fashions for gentlemen and ladies.[33] At least

[29] *The Lady's Magazine*, 1 (1770), 179.
[30] *Magazine à la Mode* (1777), 3; *The Lady's Magazine*, 1 (1770).
[31] *The Lady's Magazine*, 4 (1773), 72.
[32] *The Lady's Magazine*, 6 (1775), 235.
[33] *The Sherbourne Mercury*, 12 July 1773, Appendix.

ten other provincial papers carried fashion news on an intermittent basis as well, feeding the demand in a manner that increased the craving.[34]

Later, in the 1780s and 1790s, the London newspapers also included periodic references to dress in their reporting. By that time dress was changing so rapidly and regularly that there was an abundance of news from which to choose.[35] Usually these comments were included in the columns devoted to the social calendar of the court and aristocracy, including synopses of costumes observed at events like the earl of Salisbury's Supper and Ball, for example, or the Drawing Room on the King's birthday.[36] On occasion such accounts did double duty as critiques of the newest extravagance in dress. The editor of *The Public Advertiser* offered his readers a personal interpretation of the influence of the ballooning craze on current dress styles. 'The ballooning handkerchief is much worn: The appearance of it is promising, and balloonifises the bosom, to a magnitude, at least capable of raising one person from the earth!'[37] Whether in praise, criticism, or satire, notices of the London fads and fancies spread around the country with the delivery of newspapers. Through newspapers and periodicals, local and national, all aspects of the contemporary vogues were described to the smallest detail, with regular illustrations. The standardization of dress was more firmly established as a result.

The Barbara Johnson Sample Book exemplifies the preoccupation of members of the middle classes with the fashions of the age, while simultaneously emphasizing the effectiveness of printed sources of information. Barbara Johnson lived outside the main hub of fashion for most of her life, yet she participated in the styles and patterns of dress of the age. Visits to the metropolis and the homes of more affluent friends and relatives permitted personal observations of the *haute ton*.[38] Whatever other sources of information Barbara Johnson had, there is no doubt that printed illustrations culled from magazines and periodicals kept her perpetually in touch with metropolitan trends.

---

[34] McKendrick *et al.*, 91–2.
[35] Mrs Philip Lybbe Powys noted in her diary in 1785 that 'common ... [fashions] change every month'. Emily J. Climenson (ed.), *Passages from the Diaries of Mrs. Philip Lybbe Powys, 1756 to 1808* (1899), 220.
[36] *The Public Advertiser*, 9/11 Jan. 1787, 25 June 1787.
[37] Ibid., 16 Oct. 1784.
[38] Natalie Rothstein (ed.), *Barbara Johnson's Album of Fashions and Fabrics* (1987), 9–14.

The pages of the Sample Book contained not only the clothing fabrics bought for over sixty years, but also preserved the illustrations of dress that caught her fancy. These prints were evidently of great importance to her, conveying in the simple line-drawings the latest intelligence on current fashions in gowns, trimmings, and accessories. The first pictures depicted the costume for 1757 and 1758 and were the sort of popular engraving that appeared in the frontispiece of pocket-books. At times the clothing illustrated was clearly incompatible with Johnson's style of living, such as the plate of court dress for 1768.[39] Nevertheless, these prints were not omitted from Barbara Johnson's collection and her persistence in preserving them attests to the strong fascination fashion held, no matter how distant from her immediate experience the dresses might be; a telling commentary on the potent attraction of fashion information.

Further on into the last half of the century the fashion-prints reflected styles of clothing more within the practical realization of Johnson. In 1770 she included a plate of a morning gown and afternoon dress, while in 1771 an undress was pictured, and head-dresses were featured for a number of years.[40] More and more fashion-plates appeared in the Sample Book as the years continued. Almost every page boasted at least one or two printed examples of head-dresses, undress, and 'Dresses of the Year', garnered from many sources. *The English Ladies Pocket Companion*, *Lane's Pocket Book*, and the *New Royal Pocket Companion* supplied several of the prints, as did women's magazines and other journals to which Johnson had access. One of her brothers later moved to Kensington, and perhaps her visits there enabled her to augment her supplies of reading materials on fashion as well as the clothes themselves.[41] These illustrations form the most obvious and enduring link between modest consumers such as Miss Johnson and the fount of British fashions in London. The authority exerted by such prints and their importance to the host of consumers throughout the country cannot be over-emphasized.

Fashion-prints focused the attention of the consumers on a particular product, piquing the curiosity and sparking emulation of the ideal. The prints presented the possible and the desirable in material form; the latest styles, now within the reader's capability to imitate. More than in any previous time, this period saw fashion step down

---

[39] T219–1973, The Barbara Johnson Sample Book, Textile Department, Victoria & Albert Museum, 6–17, 19–27.
[40] Ibid. 14–17. [41] Ibid. 35–8, 40.

from the pinnacle to become a shared attraction for much of society. Modes for the middle ranks and even more humble labouring classes became eminently possible. The propagation of cheap printed information on the newest styles marked this move towards generalized consumerism among almost all classes of society.

Provincial newspapers and ladies' magazines were not the only proponents of fashion among the range of periodicals for sale by Britain's booksellers. A satirist had quipped in *The Lady's Magazine* that there ought to be an express journal devoted solely to the vagaries of fashion, the object of this implausible tract being that

> no gentleman or lady who may live within the bill of mortality, need appear on a Sunday either in church, at a park, or at a private visit, with the least deviation from the pink of the mode; if they would on the Saturday peruse the lucubrations and intelligence of this judicious and indefatigable journalist.[42]

However far-fetched this waggish proposal seemed in 1774, three years later a periodical appeared the scope of which was bounded by the dictates of fashion. The *Magazine à la Mode* declared its aim was 'to convey early and useful information to those who are in any respect concerned in furnishing Articles of Dress, either in Town or Country'.[43] As well as informing those in trade of alterations in fashions, the magazine was also intended to appeal to a general readership. Prospective readers were promised in the initial advertisements that this journal was 'Adapted to the use of all Ranks of People', and that

> In this Performance will be given an Account of the reigning Modes of Dress both for Gentlemen and Ladies. The Materials of which the Dresses are Made, will be properly explained; and the earliest Notice ... given of the minutest Alterations adopted by the fashionable world ...[44]

A co-ordinated investigation and illustration of the newest styles superseded the former haphazard selection of costume dependent on whatever caught the engraver's eye. The process of reportage and dissemination of fashion news had achieved a higher degree of professionalism, to relay more rapidly the current caprices of fashion to those hungry for news.

The literary content of the magazine no doubt attracted many of the readers outside the trade: however, to the thousands of milliners,

---

[42] *The Lady's Magazine*, 5 (1774), 540.
[43] *The Magazine à la Mode*, Jan. 1777, 1.
[44] *The Public Ledger*, 1 Apr. 1777.

mantua-makers, mercers, and tailors the detailed and timely reports would have most consequence. The hope of the publishers, and most probably its readers in trade as well, was that the barrier of distance and social accessibility could be breached; that, with the help of careful illustrations and explicit descriptions, these clothes could be reproduced anywhere.

As the time-lag shortened in the reproduction of current metropolitan modes in the newspapers and periodicals, the national market for clothing was further consolidated and the demand increasingly standardized. The tastes of consumers and the expectations with which they entered the market conformed more and more to the ideal. The outpouring of news further fuelled the textile industries of Britain. In the editorial of the first issue of the *Magazine à la Mode*, the editor acknowledged the link between promulgating fashions and promoting British industry, to the satisfaction of both parties. The editor asserted that,

every variation of the fashions gives new life to trade, both in town and country: and in the latter, an early correct communication of the true fashions, will be particularly beneficial to our manufacturers, for the traiterous [*sic*] propensity to French Manufacturers will not there prevail, as it does among the guilty great of London; it will be sufficient to have dresses made up in the reigning mode of British Materials.[45]

A ready example of fashionable dress was all the stimulus middling orders required to fire the aspiration for like costumes. Once the particulars of the style had been spread, the urge to buy an approximation invariably followed, as Barbara Johnson's scrap-book so graphically portrayed. In the circuitous round of escalating consumer demand for clothing, of prime importance was the provision of information, stimulating and channelling both interest and acquisitive instincts.

The *Magazine à la Mode* was followed by other British fashion journals, such as *The Gallery of Fashion*. Several French periodicals were also influential. In addition, women's periodicals like the *Lady's Monthly Museum* paid regular attention to the contemporary vogues in a feature called 'Cabinet of Fashion', which highlighted coloured engravings, plus detailed descriptions of the costumes illustrated.[46] Coloured illustrations became generally available after 1780, and the

[45] *The Magazine à la Mode*, Jan. 1777, 4.
[46] *Lady's Monthly Museum*, 1 (1798), 60.

whole business of presenting appealing displays of apparel became more and more sophisticated.[47] However, while the publishers of these magazines presented an ideal ensemble for the approbation of their readers, they were in fact selling only the idea of the self-striped muslin gown or the striped silk apron. The most desirable end to all this circulation of fashion news was not that several thousand daughters of the country gentry should be attired in emulation of their London cousins, but rather that many hundreds of thousands of women from the middle ranks and labouring classes should purchase gowns or items of dress in a general approximation of the latest styles.

The development of the cotton industry progressed in concert with the expanded influence of fashion in the domestic market. The products of this industry were a vast and growing fountain of fabrics—checked, striped, printed, and dyed—from low to high quality with comparable prices, made to clothe, not the leaders, but all the other degrees of which British society was composed. The spread of fashion's influence parallelled the simultaneous evolution of a profoundly different pattern of clothing-consumption than had hitherto obtained. Joan Thirsk has described in *Economic Policy and Projects* the onset of trade in clothing accessories of varying qualities for a market in Britain among the common people over one-and-a-half centuries earlier.[48] From these examples it can be seen that there were precursors to the later eighteenth-century trade; however, it was only in the last half of the eighteenth century that there was a qualitative transformation in the tempo as well as in the national character of the market for clothing. What will follow is an assessment of the alteration in the manner of buying clothes and the sorts of clothing bought in early industrial Britain.

FASHIONS IN CLOTHING: SECOND-HAND AND READY-MADE[49]

Clothing was a valuable commodity. Traditionally the fabric itself held greatest value, its utility being sufficient to encourage families to pass

[47] *Lady's Monthly Museum*, 1 (1798), 144; 2 (1799), 228; *Hollar to Heideloff*, 21–30.
[48] Thirsk, *Economic Policy and Projects* (1979), 3, 7–9, 106–32.
[49] This area is currently undergoing more extensive research than it has received to date. Clothing was not only valuable, it was also easily negotiable. The economic and social significance of clothing in the functioning of households and the wider community is being assessed in a major study of the second-hand clothes trade undertaken by this author. In addition, a closer examination of the ready-made clothing industry should reveal salient features about the place of clothing in social and commercial interaction.

garments from mother to daughter, father to son. As the eighteenth century progressed an element of selectivity entered the bequeathing of clothes. The Revd James Woodforde noted in his diary that he had given his niece Nancy 'an old brown silk gown very good never the less, and was my late Aunt Parrs'. Nancy later had this gown trimmed with fur to very good effect.[50] But such gifts of clothing no longer constituted a major source of 'new' clothing for much of the middling and labouring ranks, as once it might have done. Clothing was readily sold and replaced. The sale of clothing often constituted a source of ready income for those temporarily in need of funds, like the young James Boswell, for relative worth remained fairly constant. At the end of December 1762 he found himself short of funds, but noted in his diary:

This day I cast my eye on my old laced hat, which I saw would raise me a small supply. No sooner thought than done. Off it went with my sharp penknife. I carried it [the lace] to a jeweller's in Picadilly and sold it for 6s. 6d., which was a great cause of joy to me.[51]

Shortly thereafter Boswell recorded that he 'received for a suit of old clothes 11s.', which came in time to support the further exigencies of his style of living.[52] Another Londoner, with less demanding tastes and expenses, was doubtless furnished with a comely addition to his wardrobe from that which Boswell had sold. The trade in clothes and the movement of items of dress through society and through the market was a salient feature of pre-industrial and early industrial Britain, providing an element of choice to a greater portion of the population than has been recognized to date.[53]

Pawnbrokers, clothes salesmen, and dealers of a general sort worked in rural and urban settings, buying, trading, and selling clothes of all sorts, but looking in particular for the type of clothing they knew would be most in demand by their customers—that is, clothing in the closest proximity to the latest fashions. The discussion of the second-hand trade above disclosed how ubiquitous these traders were.[54] Other

---

[50] James Woodforde, *The Diary of a Country Parson, 1758–1802*, ed. John Beresford (1924), 41, 46.
[51] Frederick A. Pottle (ed.), *Boswell's London Journal, 1762–1763* (1950), 109.
[52] Ibid. 115.
[53] See B. Lemire, 'Consumerism in Pre-Industrial and Early Industrial England: the Second-hand Clothes Trade' *Journal of British Studies*, 27 (1988); and Lemire, 'Peddling Fashion . . .', *Textile History* 22, 1 (1991).
[54] See Chapter 2 above.

aspects of the trade are equally worthy of note. The second-hand trade was a critical factor enabling a large portion of the population to buy more apparel. The percentage of income spent on clothing would be more flexible when a part of the cost of a suit of clothes, a gown, or accessories could be recouped from the resale of old clothes. Thus, a proportionately greater access to relatively more-fashionable clothes was possible, modifying dress as the mood or the style demanded.

A further deduction to be drawn from the existence of a large, widespread market for second-hand clothes in Britain is that people of that period were accustomed to buying garments already made-up. This is an notable point, one which has to date been underrated or gone unnoticed in calculations of consumer habits. A too-common impression of the clothing trade in the eighteenth century is that there was little option but to buy yards of fabric and either make the garment or have someone else sew the item to specification. No doubt the monied classes had all their clothes made to their very specific tastes, at times taking great pleasure in unwarranted criticism of the garment in order to emphasize the singularity of their sensibilities.[55] This general pattern of purchasing clothes has been too uncritically attributed to the bulk of the population, an attribution that has never been tested and appears far from accurate. Clothing was not always obtained through the purchase first of yard goods, followed by tailoring at home or in the shop. Customers at the highest end of the scale followed this practice. But the largest and more diverse group was composed of the families of small merchants, professionals, tradesmen, innkeepers, shopkeepers, tenant-farmers, artisans, and wage-earners, none of whom was afflicted with the excesses of taste which characterized members of the higher orders. It is the general pattern of consumption of this largest group and the influence of fashion on their practices as active consumers that is of primary concern.

According to the findings of Sir Frederick Eden, writing at the end of the eighteenth century, only in the most northerly counties and parts of Scotland did one find labourers' families in which a substantial portion of the family's clothing was made at home.[56] This may be attributed more to a restricted market than to any predilection for needlework on the part of male and female inhabitants of those areas.

---

[55] The sort of behaviour that Francis Place, for one, complained of in his diary: Mary Thale (ed.), *The Autobiography of Francis Place* (1972), 216.

[56] Sir Frederick Eden, *The State of the Poor*, ed. and abridged by A. G. L. Rogers (1928), 109.

In central and southern England consumer behaviour had already altered qualitatively from that of their northern neighbours, or indeed from the customs of their grandparents and great-grandparents generations earlier. Shopkeepers now supplied nearly all clothing wants. Eden confirmed that: 'In the Midlands and Southern Counties, the labourer in general purchases a very considerable portion, if not the whole, of his clothes from the shopkeeper.'[57] Some of the clothing worn by all but the poorest people would have been made to order, purchased from local tailors or drapers. Yet it is also true that an indeterminate part of the clothes bought in Britain were bought already made-up, either second-hand or ready-made.

With few exceptions it has been assumed that ready-made clothes were rare until later in the nineteenth century.[58] The one history of the manufacture of women's clothing states that:

There was little production of ready-made outer clothing for women. The earliest manufacturers were retail shopkeepers. The department stores, in the second half of the nineteenth century, in order to keep their workroom staff occupied at times when they had a small demand for bespoke garments, began to make rather simpler garments for sale ready-made. Other pioneer manufacturers began, in the 1890s and early 1900s, to produce aprons, mob caps, sun bonnets, frilling, blouses, overalls and cloaks.[59]

Another more recent historian dated the appearance of ready-made clothes for the working classes to the same period.[60] Ready-made clothing was available much earlier than might be supposed from these statements. The first organization to sell ready-made clothing in Britain had a vast production unit, materials were cheap, and production was designed to be low-cost, appealing to the widest market. The commercial characteristics of this enterprise have the hallmark of a modern manufacturing company; however, the agency

---

[57] Ibid. 108.
[58] The greatest amount of research has been given to the recording of fashions during this period. However, more recently there has been a shift of focus which to some extent developed out of work on the production of textiles and associated goods in the eighteenth century and earlier, Stanley Chapman and Joan Thirsk being two of the earliest working in these areas. See also B. Lemire, 'Popular Fashion and the Ready-made Clothes Trade, 1750–1800', *Textile History*, 15 (1984), and the forthcoming volume of *Textile History* on the economic and social history of dress. See also Carole Shammas, *The Pre-Industrial Consumer in England and America* (1990), chapters 4 and 6.
[59] Margaret Wray, *The Women's Outerwear Industry* (1957), 18.
[60] W. Hamish Fraser, *The Coming of the Mass Market, 1850–1914* (1981), 101.

was in actuality the East India Company, headed by Sir Josiah Child. Late in the seventeenth century Child sought to diversify the products of the Company in order to attract a wider range of customers. To this end he began the first large-scale sale of ready-made goods in England. Child was concerned about a possible over-abundance of Indian fabrics in Britain, and worried about a consequent drop in the price of Indian cottons. Accordingly, the dispatched written orders to Fort St George on the Coramandel coast, requesting that the factors there arrange for calicoes 'to be strongly and substantially sewed for poor people's wear', in order to attempt an unprecedented substitution of cotton for linen in 'the wearing of calico in shifts'.[61] The bulk of the cotton shifts imported were made of strong blue-and-white fabric suitable for seamen and labourers and there were also two other sorts of shifts made. Some of middling quality were thought suitable for tradesmen and others of their ilk; while fine white cotton was carefully sewn in shifts designed for ladies and gentlemen. Child specified that 200,000 shifts in all be manufactured for sale in Britain and other parts of northern Europe.[62] This marked the first marketing on such a massive scale of ready-made clothing: the level of production makes this case unique. None the less, the fact that Child considered this initiative at all suggests that ready-made goods were known in his day, and in the decades that followed there is abundant evidence of an extensive and increasingly diverse quantity of such clothing.

Prior to the eighteenth century, cottage manufacture made several sorts of ready-made apparel available to consumers. Joan Thirsk has documented and described the onset of the trade in ready-made accessories such as knitted stockings nearly two centuries earlier.[63] There was no diminution of this trade. Many more items than simple accessories were obtainable. More recently, Margaret Spufford has considered the availability of merchandise including ready-made clothes in *The Great Reclothing of Rural England*. In the records of six men from Kent, described as 'salesmen', there are lists of many sorts of light cotton and linen clothing such as shirts, drawers, petticoats, and frocks, which they sold between 1680 and 1721. In addition, they supplied heavier garments among which were ready-made gowns of

[61] Dispatches from England 1681 to 1686 (Fort St George), 9 Oct. 1682, p. 15, quoted in K. N. Chaudhuri, *The Trading World of Asia and the English East Indian Company* (1978), 287. [62] Chaudhuri, 287.

[63] Thirsk, *Economic Policy and Projects*, 6, 8; id., 'The Fantastical Folly of Fashion; the English Stocking Knitting Industry, 1500–1700' in K. G. Ponting and N. B. Harte (eds.), *Textile History and Economic History* (1973).

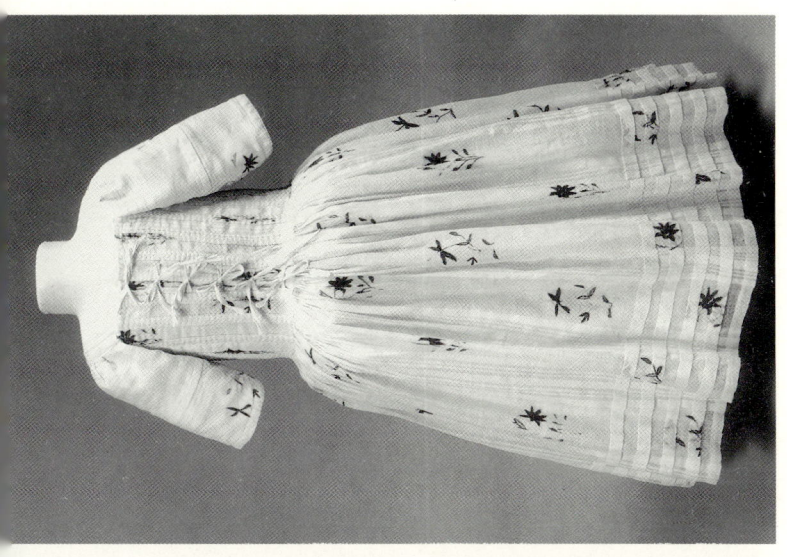

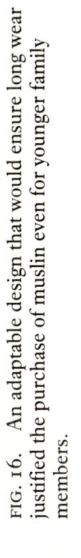

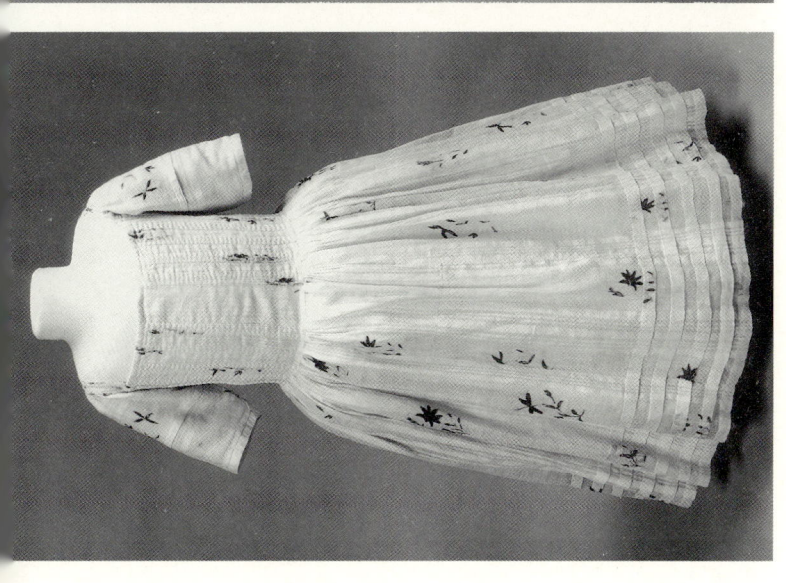

FIG. 15. As the price of muslins fell fashionable muslins were used even for children's clothing.

FIG. 16. An adaptable design that would ensure long wear justified the purchase of muslin even for younger family members.

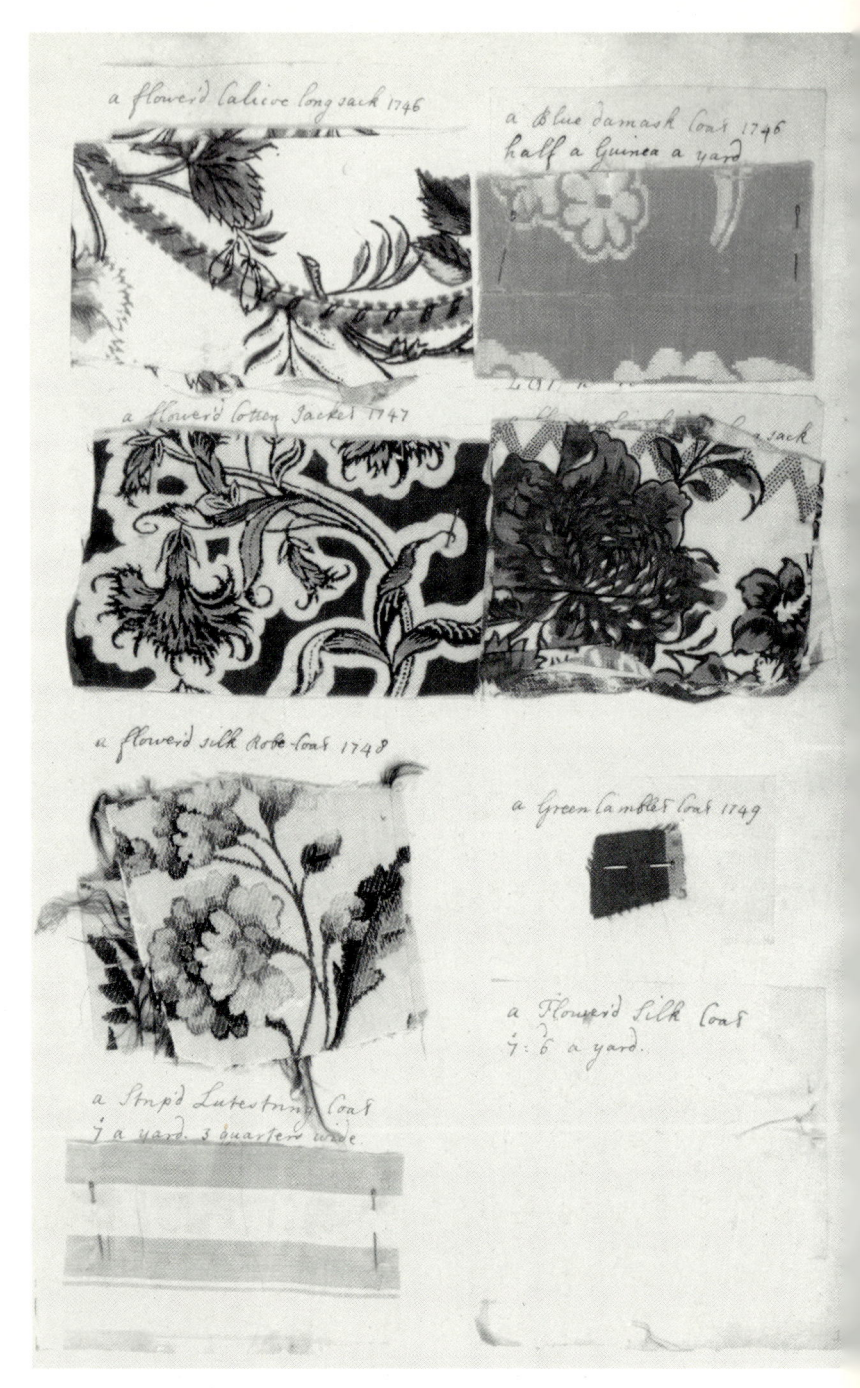

FIG. 17. Barbara Johnson's sample book reflects a gradual shift in her purchases from predominantly silks from the 1750s to 1770s, to a greater concentration of cottons thereafter.

FIG. 18. Greenfield's trade-card displays his trade in knitted and woven cottons of British manufacture.

FIG. 22. Illustrations in women's magazines supplemented formal fashion plates, providing additional idealized glimpses into the fashionable world.

FIG. 23. The proprietor of this gown warehouse emphasized the specialization of his trade and the variety available to his customers.

MAGAZIN·DES·ROBBS·DE·CHAMBRE

In the Temple Exchange Coffee House Paſſage, over-againſt St. Dunſtan's Church.

Is opened a new Gown Warehouſe, with a clean, freſh Stock of Morning Gowns and Banjans of Scotch Plods, Silk Damasks, Stuff Demasks, Flower'd Silks, Thread Sattins, Turkey Mantua's, Floretta's, Callimancoes, &c.

As Gowns are the intire Trade of this Warehouſe, all Gentlemen may depend upon good Choice, and being ſerv'd at very reaſonable Rates. 1732

FIG. 24. Shopkeepers, like Mary and Ann Hogarth, provided an array of inexpensive ready-made apparel made of hardwearing fabrics.

FIG. 25. Clothes dealers like John Keet operated in major cities and ports, preparing clothing in quantity for domestic and overseas markets.

FIG. 26. Allin sold new and used clothes at low prices, aiming to please the Birmingham consumer. He was part of a complex contingent of dealers who served provincial Britain.

worsted fabric and damask mantuas, as well as many sorts of breeches. There was no indication whether these items were new or second-hand; however, the ready access of chapmen to these sorts of items and the presence of these goods in their bales implies a familiarity among both salesmen and customers. Dr Spufford speculated that these tradesmen may well have been the precursors of later tradesmen in ready-made goods, a supposition that has been borne out. Clothing was available ready-made before the opening of the century, and became more and more common as the century progressed.[64]

Aside from hats and hose, ready-made clothing of different sorts was sold in London and the provinces. Evidence from metropolitan shops is more plentiful where a range of sometimes costly items are recorded. Banyans, or morning gowns, could be bought ready-made from numerous retail premises for men, women, and children. An advertisement for this type of garment appeared in 1718 for the 'Original Gown-ware-house at Baker's Coffee-house Exchange alley, Cornhill', while a later trade of the same type was noted at 'Harrison's Ware-house against the King on Horseback, Charing Cross'.[65] (See Plate 23, an advertisement for a gown warehouse at Temple Exchange.) At these establishments and many more like them, ready-made gowns for lounging at home were sold in the most fashionable materials. The earlier advertisement announced that gowns for 'Men, Women, Boys, Girls, and Children in Arms, are continued to be sold by Wholesale and Retail'.[66] These were offered in addition to the caps, sashes, and quilted petticoats which were also stocked. The gowns were made up in many sorts of silks and fine worsted, and prior to 1721 in calico as well. In 1733 ladies' gowns 'with Silk Waddings' were on hand for the London shopper, as too were striped and plain mantuas. By the 1730s, at least five London gown warehouses specialized in morning gowns, their trade characterized by relatively expensive ready-made clothing intended for the moneyed classes, not for common consumption. Others, such as hosier John Hill, offered to the less affluent items such as 'Women's Cotton, and Worsted Pettycoats, and Waistcoats' at the same period.[67]

How common these traders were has not been finally determined, but they were not a rare or exotic sight. William Hogarth's two sisters, Mary and Ann, opened a shop about this time, for which the noted

[64] Margaret Spufford, *The Great Reclothing of Rural England* (1984), 123–5.
[65] *The Post-Boy*, 26 Feb. 1718; *The Country Journal or The Craftsman*, 3 Aug. 1728.
[66] *The Post-Boy*, 26 Feb. 1718. [67] *The Country Journal*, 23 Mar. 1734.

engraver prepared a trade-card to advertise their wares. The shop was known as the 'Frock Shop', and the card featured a very attractive interior view, showing a young boy being fitted with a frock coat in the company of his family. Ready-made clothes hung from the wall and were piled on the counter. The sisters announced that

> Mary and Ann Hogarth ... Sells ye best & most Fashionable Ready Made Frocks, suites of Fustian, Ticken & Holland, stript Dimmity & Flanel Wastcoats, blue & canvas Frocks & bluecoat Boys Dra'rs, Likewise Fustians, Tickens ... in ye piece, by Wholesale or Retale, at Reasonable Rates.[68]

(See Plate 24, the Hogarth trade-card.) The compiler of Hogarths' engravings describe the business as 'a line draper's, or rather, what is called a slop shop', showing a casual attitude to the designation of the business that is frustrating to historians, but most probably typical of an era when the differentiation in these sorts of trades was not distinct.

Shopkeepers of many sorts would have needed some sorts of goods ready-made. Handkerchiefs were the most basic article in this category, and at the same time were an essential part of the dress of the eighteenth-century woman. Tucked around the neck or draped round the shoulder, it was an accessory used continually during this period. At the least a handkerchief required hemming, a process easily mastered by even an indifferent seamstress. Handkerchiefs may have been made up by the mercer's employees; but they could just as easily have been hemmed by members of the weavers' families or by women employed by a merchant, much in the way that caps and stockings were produced by part-time domestic workers.[69] Thomas Mortimer notes in his 1763 edition of *The Universal Directory* for London, that 'the common Morning Cap for men are sold by most Haberdashers, requiring little skill or ingenuity in making, and being generally performed by poor women'.[70]

Garret seamstresses working by candlelight for a pittance attracted the notice of reformers in the nineteenth century. Thomas Hood's 'Song of the Shirt', published in *Punch* in 1843, was a response to an earlier article on the distress of seamstresses employed in piece-work production of ready-made clothing.[71] The attention paid this sort of

---

[68] *The Complete Works of William Hogarth in a Series of One Hundred and Fifty-seven Engravings, From the Original Pictures ... with Descriptions and Comments on their moral Tendency ... by Rev. John Trusler. Accompanied with the Complete Life and Numerous Anecdotes of the Author and his Works by J. Hogarth, Ireland and Others* (n.d.), ii, 275.

[69] Thirsk, *Economic Policy and Projects*, 6–8. [70] *The Universal Directory*, 14.

[71] Gertrude Himmelfarb, *The Idea of Poverty* (1984), 313.

work in the 1840s reflects more the reforming environment of the period than the sudden emergence of this type of employment. Sweated labour did not originate in the nineteenth century. Its appreciable expansion was probably the result of the worst sort of putting-out system, which flourished in conjunction with the rising output of textiles, growth of population, and demand for ready-made wares. The 'discovery' of this sort of misery awaited the zeal of investigators in the reforming age. No eighteenth-century reformers scoured the rookeries and courtyards of London to leave records of the precursors to the nineteenth-century sweated trades, but this sort of employment was the resort of the poor in the eighteenth century as it would continue to be in the nineteenth.[72] As the textile trades flourished, the making of clothing by piece-work developed as an adjunct. Francis Place and his young wife Elizabeth worked together in their room sixteen to eighteen hours a day in exactly this sort of manner, sewing breeches for a tradesmen when no other work was available. In spite of the hours, at the time they were glad to have the work as insurance against starvation.[73] Aprons, caps, ruffles, and pockets were equally commonplace ready-made retail items. None of these commodities needed a great deal of skill to make, and it is not surprising to find them listed repeatedly in records from the early part of the century.[74] Needle and thread could as easily be put to use in producing ready-made goods, as could shuttle and yarn in the creation of yard goods. When presented with the evidence from shops in London, Kent, Northampton, and Chester, in addition to the contents of chapmen's bundles, there can be no doubt that articles of dress were prepared beforehand to be ready for customers who wanted clothes immediately and cheaply, and did not require custom-made merchandise. Simpler sorts of apparel and accessories were manufactured for general consumption: this had been the case for some time. The more complex articles of clothing began to appear in the

---

[72] An example of this sort of humble needlework appears in the 1715 document, *The Ordinary of Newgate His Account of the behaviour, confessions and Dying words . . .*, 23 Dec. 1715, 5, in which Martha Pillah related her pattern of employment. 'She said, she was about 18 years of age, born of very honest Parents in Brewers-yard, in the Parish of St Margaret, Westminster: That her Friends put her out Apprentice to a Taylor, and when the Time of her Service with him was expir'd, she work'd for herself, whose chief Business then was, the making and mending Men's Cloaths'.

[73] *The Autobiography of Francis Place*, 123.

[74] The 1740 inventory of a bankrupt draper in the Northampton area listed fifty-seven handkerchiefs and two-dozen caps among his stock. YZ/8366, Northampton Record Office.

inventory of individual shops as early as the late seventeenth century and on into the eighteenth century, surfacing with greater frequency as the century progressed, as consumer custom adapted, and as the market expanded.

The East India Company stands out as the largest purveyor of ready-made clothing to date. None the less, it was not the only producer on a large scale either during the seventeenth or the eighteenth century. Manufacture of clothing for the military took place throughout this period, but since in the case of the land-forces there were specific uniforms, and not ordinary working garb, they have not been of interest to historians dealing with the question of general clothes production. Clothing for the navy, for the common sailor, was different and is of direct relevance to the topic of ready-made clothes. Sailors' dress was not uniform. Only in 1663 were the most basic injunctions in the standard of dress issued by the duke of York, that the 'Monmouth caps, read caps, yarn stockings, Irish stockings, blue shirts, white shirts, cotton waistcoats, blue neckclothes, canvas suits and rugs, . . . are alone permitted to be sold'.[75] Extra supplies of clothing would be stocked on board ship in slop chests, so called after the name 'slops' given the wide-kneed breeches worn by sailors. These supplies would then be sold as required during the voyage. Merchants were contracted to supply various of the cheap, ready-made garments, and with the process emerged the first large-scale production of common, ready-made apparel. The size of the orders were vast. Bales of breeches, shirts, and jackets, were conveyed from the site of manufacture to the principal naval ports.[76] The importance of this sort of clothing does not end with the navy. The garments were made in the style of working-men's clothing, therefore the market extended beyond the service's needs. So successful was the production of slops that this type of clothing was used extensively by different segments of the population, some willingly and others perforce. People condemned to be transported to Australia, for example, began their ordeal by being stripped of their own clothing and given slops in exchange.[77] The word

[75] Admiral Sir Gerald Dickens, *The Dress of British Sailors* (National Maritime Museum Publication, 1977), 4.

[76] Adm. 49/35; Adm. 106/2584, Public Record Office.

[77] I am indebted to Margaret Maynard, lecturer, Department of Fine Art, University of Queensland, for the information on the use of slops in transportation. Robert Hughes, in *The Fatal Shore*, paints a macabre picture of the hapless convicts being stripped of all valuable pieces of clothing by the captains of the transport ships prior to embarkation. The clothes were then sold on shore and must have represented a flourishing part of the local trade in second-hand clothes. For many of the 'respectable' convicts this was the

'slop' entered the language as a description of sailors' loose trousers, but shortly thereafter came to mean 'ready-made, cheap or inferior garments, generally'.[78] Just so, the slops manufactured in Britain ceased to be exclusively nautical gear and became part of the wear of the labouring ranks wherever shops were established.

A number of circumstances combined in the eighteenth century to encourage and promote the manufacture of ready-made clothing in Britain. The first of these factors was the extraordinary diffusion of fashion throughout Britain, both geographically and socially. This phenomenon tied consumers of clothing to one impulse, which, as it radiated out, commanded all consumers to some degree. The impact of the rapid dissemination of styles through national publications centralized production of fabrics, while an improved distribution-network resulted in unparalleled uniformity in tastes and consumption. Many consumers were accustomed to buying some clothes ready-to-wear, yet at the same time had at least some inclination to fashion. Among this population expenditures would vary. However, aspirations to fashionability could be fulfilled to some degree with only small expenditures on the elements of dress. Cotton textile products solved the dilemma of the British consumer, both for the small purchases and larger, more regular additions. The success of the cheap, British-made products mirrored the earlier achievement of the East India Company; and like the East India Company, British entrepreneurs realized the advantages of diversifying their commodities.

The development of large-scale production of ready-made gowns and other apparel grew with the expansion of the cotton industry. In the late 1750s invoices of orders from Lancashire reveal several sorts of ready-made clothing. One order requested five types of breeches, in different colours. Caps were also ordered: two seam caps, printed with flowers, or scarlet, of several qualities, while the caps ordered the following year were coloured predominantly blue and white.[79] These sorts of goods regularly appeared in invoices during the mid-century.[80]

---

beginning of a degradation the like of which they had never imagined, beginning with the loss of their respectable appearance, their clothes. *The Fatal Shore* (1987), 139.

[78] *The Shorter Oxford English Dictionary*, ed. C. T. Onions (1973), ii, 2,021.

[79] Invoices, Peach & Pierce, 12 Mar. 1757, 12 Apr. 1758. Beekman Papers, New York Historical Society, New York.

[80] Joan Thirsk discussed, among other products, the cottage production of knitted caps, felt hats, and Monmouth caps in the seventeenth century for both home and export markets, *Economic Policy and Projects*, 2, 115, 129. Experience of this sort prepared a work-force for the requirements of clothing production over the next century.

The work involved in these simply made garments allowed for ease of manufacture, and even in more complex items like breeches, after the skilled cutting the seams could be set with minimal training.[81]

Little documentation remains concerning the merchants or tradesmen who undertook the manufacture of this sort of clothing. However, there is some mention of the Bristol merchant John Peach, who described his trade as follows: 'Besides Irish Linens I deal largely in Irish Sheeting Brown and White and Dyapers of all Breadths, Brown Hempen and Flaxen Sprigg Linen and Hessen and Holles. Likewise printed Linen, Cotton and Printed Hats'.[82] The trade in linens and cottons appears to have been naturally extended to include accessories of the same fabrics. Peach acted as the putter-out, sending the materials to out-workers to be made into headgear, after which he arranged the sale of the merchandise, sometimes providing his relative Samuel Peach with articles for his overseas trade. John Peach exemplified what was probably a common arrangement among the tradesmen and merchants large and small, whereby pieces of clothing were manufactured in large quantities for an equally large market. The size of this sort of enterprises is suggested in the volume of caps supplied for one order. In 1757, Peach & Pierce sent one customer sixty-dozen caps of ten different colours, in one consignment. The orders for handkerchiefs matched those for caps.[83] Altogether, the trade in ready-made clothing comprised a small portion of the cotton trade at this time, but its close connection to the source of fabric production showed that manufacturers and traders were ready and willing to diversify their products where they perceived the potential for profit.[84]

By the late 1760s there was a growing line of women's ready-made clothing associated with the manufacturers and traders of British cottons. The ready-made industry of this period grew up beside and in conjunction with the Lancashire cotton industry.[85] Ready-made

---

[81] *The Autobiography of Francis Place*, 95.
[82] Philip L. White (ed.), *The Beekman Mercantile Papers* (1956), ii, 662–3.
[83] Invoices, Peach & Pierce, 29 Mar. 1757, Beekman Papers.
[84] See footnote 127 below, John Totterdell of Bath, for an example of another clothing manufacturer, associated with a region of textile production.
[85] Note in the table of the Manchester trades in Wadsworth and Mann, 257, the presence of five petticoat manufacturers in the list of producers for 1781. This can be seen as a progression consistent with other specialized types of production, such as hat and fringe manufacturers, who are also listed at this time.

products affiliated with the cotton industry point to the recognition by manufacturers of the homogeneity of fashion and the further potential within that market that could be cultivated. The first orders for ready-made gowns associated with the cotton industry were found in 1767. At that time Beekman requested '12 fine Cotton Gowns' in 'Handsome Colours' from Robert and Nathaniel Hyde, Manchester check manufacturers and warehousemen. The gowns were priced at nine shillings each.[86] References to ready-made gowns recurred in other orders sent to the Hydes; however, orders from the same customer sent to Bristol never contained orders for cotton gowns.[87] A reasonable inference might be that the manufacture of cotton gowns was centred in the Manchester region, not in Bristol. The skill needed to tackle the creation of a basic gown would not exceed the ability of even a moderate seamstress, for the large running-slip stitch was the more common sewing style than all-but invisible stitches. This level of skill was in abundance, as too were the expanding varieties of fabric. Thus, it was to Manchester that orders were sent for gowns of many colours, patterns, and fabrics, made up in the materials of the regional textile industry.

The records of an unidentified Manchester firm reveal the most comprehensive evidence of the abundance of ready-made clothing in the later eighteenth century and the association of the cotton industry with the provision of these goods. The Day Book of sales of this firm chronicles the sale of ready-made gowns to customers throughout Britain and parts of Ireland during the years 1773 to 1779. Some of the destinations for these orders include Nottingham, Liverpool, Chester, Hull, Buxton, Chesterfield, Sligo, and London.[88] London provided the largest market for ready-made clothes of all descriptions. The firm shipped thousands of ready-made gowns to the city each year, to the many wholesalers established in its environs. One of the largest customers received over 800 gowns in a variety of colours and patterns and in several fabrics, over a one-year period from 1777 to 1778.[89]

This Manchester firm sold several sorts of gowns. The cheapest cotton gown cost customers from 8 shillings to 8 shillings and 6 pence at the start of the ledger and cost only 7 shillings and 6 pence or 7

---

[86] Invoices, Robert & Nathaniel Hyde, 12 Mar. 1767, 10 Jan. 1768, Beekman Papers.
[87] *The Beekman Mercantile Papers*, ii, 857, 861–4.
[88] MC: MS ff 657, D43, Day Book, of sales of an unknown Manchester firm, 1773–9, Manchester Public Library, 4, 280, 318, 322, 371, 444, 470.
[89] Day Book, 492–581.

shillings and 9 pence by 1779. The next in price was the silk-and-cotton gown, priced from 16 shillings to 17 shillings in 1773 and from 11 shillings and 6 pence to 15 shillings and 6 pence in 1779; while the most expensive was the moree gown, made from a cloth which could have been either a wool-and-cotton mixture or the finest cotton, with a watered finish. The moree gown sold for anything from 21 shillings to 43 shillings and occasionally as low as 16 or 17 shillings.[90] The range of prices for the moree gown reflects the refinements in dyeing and printing that the firm offered its customers. The cotton gowns were usually sold in dark and light shades, and gowns so described were the ones most commonly available. As well, cotton gowns were procurable in blue-and-white and pink-and-white patterns, plus plain pink.[91] Moree gowns came in the greatest number of patterns and colours. A figured moree gown sold for an extra shilling; one printed with darts for 2 shillings more; spots cost an extra 5 shillings. Gowns dyed either pink or red were always more costly, as these dyeing processes were lengthy, whereas an orange moree gown cost only the standard 21 shillings.[92] Ready-to-wear gowns could be had in a multitude of colours and printed patterns, and in addition the firm accepted commissions for custom-designed gowns. In June 1777 Mrs MacNeale of Buxton received her first consignment of four sorts of specially patterned gowns costing from 36 to 43 shillings each. The following month she ordered three extra gowns of this sort, and next summer a further order was placed for additional special-order gowns, including one which was to be 'Flowered in the Loom'.[93] These articles appealed to a wide cross-section of consumers. The cost of the moree gown would have restricted it to those more-affluent members of the artisan and middling classes. The cheapest cotton gown, however, cost only 7 shillings and 6 pence just slightly more than the 6 shillings and 6 pence deemed by Sir Frederick Eden an acceptable expenditure for 'a common stuff gown'.[94] The additional shilling procured an attractive, fashionable attire for the less-affluent consumer. Moreover a cotton gown had several advantages over a woollen one, being easily washed and probably aesthetically more pleasing in a colourful print. The silk and cotton gowns offered a better quality at less than a pound. The

---

[90] The author of *The Merchant's Warehouse laid open: Or the Plain Dealing Linen-Draper*, published in 1695, described moree as 'a Callico in use in Drawing of Work, for Petty-coats and waste-Coats'. It is most probable that the fabric called 'moree' in the 1770s was a facsimile of the Indian moree.
[91] Day Book, 300, 304, 315, 329. [92] Ibid. 317, 318, 340, 347, 349.
[93] Ibid. 465, 470. [94] Eden, 110.

other major advantage of this clothing was that it was all ready to wear, no major preparation being necessary.

The fashionability of the dresses was assured, made as they were in Britain and at a time when improved transportation facilitated both the transmission of fashions to the manufacturing centres and the easy distribution of the goods across the country. The Manchester manufacturers kept a close watch on all the innovations in fashion, so fabrics could be created in line with the newest styles, and novelties introduced in accordance with the reigning modes. Samuel Oldknow was advised by his London distributor Samuel Salte to 'vary the spot, barley corns, leaves and other little fancy objects'[95] so as to make the more distinctive goods and take advantage of the insatiable demand for originality. Producers paid the closest possible attention to the shifts and changes in London fashions to ensure that their products met with the approval of the metropolitan customers. Those who made ready-made gowns were equally assiduous in their study and cultivation of the metropolitan market. The success of the red-coloured moree gown is a case in point (see previous chapter's discussion of shipments made to London).[96] The aim of both manufacturers and merchants was to serve the domestic market and answer the sometimes unanticipated requirements of a popular fashion.

The Manchester firm's ledger reveals the new sorts of clothing that came on the market with each passing year. Not all of these items were ready to use or wear immediately—waistcoat shapes were just such items. The Day Book records the sale of eleven varieties of these products: 'Velveret Buff Shapes', 'Print'd Waistcoat Shap', and 'Silk Waistcoat Shapes', reflecting in those items the variable tastes of customers at whom the goods were aimed.[97] The velveret would have been hard-wearing, the printed waistcoat probably less so but more decorative, while the silk waistcoat would have been much more attractive. These articles were not made-up; they were not ready-to-wear. However, the most difficult part of converting yards of cloth to clothing had been completed in the preparation of the individual shapes. Subsequently, the shapes could be embroidered either professionally or in the home—ladies' magazines contained regular patterns for this sort of embroidery. In addition to gowns and waistcoat shapes the Manchester firm supplied the simpler sorts of accessories.

[95] Letter from S. Salte to S. Oldknow, 13 June 1786. Oldknow Papers, Eng. MSS, 751, John Rylands Library, Manchester.
[96] Day Book, 411, 418, 442, 458, 462.
[97] Ibid. 361, 543.

Standard linen and check handkerchiefs were joined by more exotic and expensive kinds. Mallabar handkerchiefs, printed after the fashion of cotton goods sent to East Africa, became a favoured article as did the silk and muslin handkerchiefs.[98] By 1776 silk cravats had been added to the list of merchandise supplied by the Manchester firm.

The products of the British cotton industry enabled a larger portion of the population to display a greater degree of fashion than ever before. The ready-to-wear clothing combined the essence of popular fashion with a range of garments that had immediate utility. The combination of popular fashion in a ready-made format became a key feature of modern manufacturing and marketing. The success of these entrepreneurs in promoting fashion trends can perhaps be measured by the complaints of luxury among the lower orders in 1783, when 'Every servant girl has her cotton gown and cotton stockings'.[99] Another contributor to *The Lady's Magazine* begged for a distinction in the fashions of the varying ranks in vain:

> Do Madame, pray let there be some alteration made in the ladies sacks, for I am quite tired of seeing every lady's woman and housekeeper drauling about with them just the same as their ladies have on, made exactly of the same taste, and sometimes in a higher ...[100]

Yet, however much some in Britain might repine over the alterations in dress of the middle and working classes, the popular adherence to fashion could not be quashed, nor could the associated commercial and industrial trends be reversed. The presence and proliferation of fashionable, ready-to-wear clothing in the eighteenth century context indicates the degree to which the diminution of visual social distinctions was an element of that period. Moreover, the marketing of stylish, inexpensive items in the mass market suggests how integral to the success of industrialization was the popularization of fashion.

RETAILING, FASHION, AND READY-MADE CLOTHES

Consumer trades in London were powered by the exigencies of fashion, as every tradesman knew.[101] The primacy of style, combined

---

[98] Day Book, 293, 339, 387.
[99] *The London Magazine*, 52 (1783), 128.
[100] *The Lady's Magazine*, Feb. 1773, 68.
[101] Tradesmen could ignore the dictates of fashion only at their peril, as Francis Place's master discovered during the former's apprenticeship. 'My master at the time I was apprenticed had nearly outlived his customers', Place recalled, 'and as the business

with the great numbers flowing through the capital, sustained the great metropolitan market for clothing of all sorts. London offered possibly the largest assortment of ready-made clothing to the eighteenth-century consumer. Shopkeepers large and small displayed all the products of Lancashire's looms and workshops, but the unique characteristics of the London market encouraged the development of an indigenous ready-made industry. The young apprentice surgeon William Clift, newly come to London from Bodmin in Cornwall, entertained his sister with written accounts of his life in the capital. A month after his arrival he described to her in some detail the amendments he had made in his clothing, commenting as well on the general availability of ready-made clothing.

I have to tell you that I have a new Suit of Clothes making and I am in hopes I shall have it to wear to Morrow it is of a Dark brown and the Coat waistcoat and Breeches are of one sort, I have three new Shirts that cost Seven shillings Each they buy them here at the Shops ready made, I have 3 Neckcloths of the same sort of them I had of Brother John and I have 4 new Pocket Handkerchiefs...[102]

While William did not yet compare his appearance to that of the metropolitan barbers and tailors who, with their wives, he described as, 'All dukes and Dutchesses on Sundays', his new attire brought him evident contentment.

Linen-draper Harry Barker of 11 Pall Mall was one of those who advertised ready-made cotton shirts in calico as well as check. The proprietors of London clothing warehouses, aiming to serve the local market and the tens of thousands who passed through the city, realized that it was to their advantage to provide certain sorts of clothing ready-made in the current mode. The large, growing work-force made this sort of undertaking probably less difficult than custom-tailoring.[103] Earlier in the century London had been the site of production of garments for the extremes of the market: bales of cheap slops and

---

was to a great extent one of fashion and was undergoing great changes which he could not adopt, so he could procure no new customers', *The Autobiography of Francis Place*, 74.

[102] Letter, 10 Mar. 1792, William Clift to Elizabeth Clift, quoted in Francis Austin, 'London Life in the 1790s', *History Today*, 28 (Nov. 1978), 741.

[103] There is also the suggestion that items are arriving ready-made from other areas of specialized textile manufacture, such as Ireland. Several ads in the 1760s announce 'a compleat Assortment of fine Holland Shirts, lately imported' from that location. This would suggest that several districts were taking advantage of the evident market for ready-made articles and providing not only the yard goods, but also the finished items themselves. Further research is required on this question.

charity clothing at one end[104] and the selective sale of costly Indian-silk morning gowns at the other. By the 1760s ready-made clothing was easily available not only at the two extremities of the market but also in the midding sector, where quality, price, and style were the determining criterion. Advertisements are sprinkled through the London papers, and the concerns of the customers are reflected in the copy. Robert Blunt of the Golden Ball, Charing Cross, promised 'SHIRTS and SHIFTS made in the neatest Manner of fine Holland, or Irish Cloth plain or ruffled, the Needlework elegantly performed. Shirts for the Army, or strong yard-wide Linen, for Families, or Gentlemen going abroad, ready-made'.[105] Mrs Roberts, at the Acorn in Salisbury Court, Fleet Street described 'a compleat Assortment of fine Holland Shirts, lately imported, from 12s. to 25s. a Shirt, and several fully-trimmed, at 14s. 15s. and 16s. neatly made of fine Irish Linnen, with all other Prices down to 5s. a Shirt plain'.[106] Forster's Linen Warehouse announced a 'great Choice of ready-made Shirts of all Prices from 4s. 6d. to a Guinea a-piece, among which are some fine and neatly made at 10s. a-piece'; while a year later the ads specified that 'A Parcel of Callicoes and Callico-Shirts are now selling very cheap'.[107] In Piccadilly, at the corner of the Haymarket, Mr Bryan, tailor, advertised the arrival of 'A Parcel of neat fine Holland Shirts, both plain and ruffled, from Fifteen to Twenty Shillings per Shirt, worth Twenty four Shillings a Shirt. Likewise a Parcel of fine Holland Shifts at the most reasonable Rates.'[108] These then were a few of the many who systematically advertised their wares through the pages of the daily newspapers, citing the attractions of their particular stock for the readers.

The large traders in ready-made clothes operated through warehouses open to the public. The attraction of these emporiums was the promise that prices would be lower and the selection wider than in smaller or more select shops. In 1763 Robert Blunt ran a warehouse for ready-made shirt at Charing Cross.[109] There were at least two

---

[104] One advertisement for this sort of product read: 'Tho. Woods, At The Indian Queen and Ship, at the Corner of Houndsditch without Bishopgate, London, SELLS all sorts of Cloaths for Charity-Schools and Workhouses, as cheap as any Body that makes them so good as he does; where Proposals may be had. Charity-School Boys Caps with Tufts and Strings, wholesale and retail', *The Daily Advertiser*, 15 Jan. 1740.
[105] *The Public Advertiser*, 4 Jan. 1766.
[106] Ibid. 19 Jan. 1763.
[107] Ibid. 15 Feb. 1764, 19 Nov. 1765.
[108] Ibid. 16 July 1763.
[109] *Universal Directory* (1763), 111.

other 'Linen and Shirt Warehouses', one at St Paul's and the other also at Charing Cross.[110] The notice for 'Bromley's Linen and Shirt Warehouse' in Charing Cross, promised 'Any Gentleman having immediate Call for ready made Shirts may be supplied with any Quantity, from 5s. 6d. to 21s. finished in the neatest manner'.[111] In addition to shirts these warehouses stocked handkerchiefs, stocks, neckcloths, and table linen. Moreover, these were only some of the speciality warehouses situated in the capital. Petticoat warehouses offered petticoats of all descriptions, in every type of fabric, plain or quilted.[112] A further specialization of the petticoat warehouse were those which sold only quilted petticoats, such as Mr Taylor's 'Quilted Coat Warehouse', the newest of which was opened in Ludgate Hill in 1779. He owned two other similar establishments, one in Covent Garden and the other in Bishopsgate Street. At the opening of the Ludgate Hill premises Taylor promised that he had 'laid in a very large assortment of fresh quilted coats, in new and elegant patterns, never before quilted', and he assured customers that 'should they not be able to suit themselves with petticoats at either of his warehouses in Covent-garden or Bishop-gate-street, they may depend on being accommodated at his warehouse upon Ludgate-hill, as much greater variety is kept there, it being a more central situation'.[113]

Besides the warehouses selling linen, shirts, and petticoats there were those which featured waistcoat shapes. One advertisement stated that their establishment sold waistcoat shapes already embroidered and offered a discount to anyone who wanted to buy a supply to prepare for the ready-made trade.

An Elegant Assortment, just finished, suitable to the Season, on rich Sattin and other Grounds, worked in Scheneels, Gold, Silver, and Colours. Likewise great choice of plain and spotted Silk Shaggs, Velvets, etc. Gentlemen going to the East or West-Indies may be supplied with great Variety of Patterns, suitable to those Climates, on proper Terms for Exportation. Good Allowance to the Trade and those who make them up.[114]

The catch-phrases of all the advertisements were the same, promising clothes 'neatly' made, in an 'elegant assortment', or newest pattern,

[110] *The Gazetteer and New Daily Advertiser*, 8 Jan. 1778; *The Public Advertiser*, 14 Jan. 1771.
[111] *The Public Advertiser*, 14 Jan. 1771.
[112] *The Public Ledger*, 1 Apr. 1771; *The Gazetteer and New Daily Advertiser*, 14 Jan. 1779.
[113] *The Gazetteer and New Daily Advertiser*, 14 Jan. 1779, 12 Oct. 1779.
[114] *The Public Advertiser*, 16 Jan. 1771.

recognizing the necessity of offering not only the current fabrics, but also clothing in the contemporary mode. Establishments like Mr Bryant's 'Great Coat and Mecklenburgh Cloak Warehouse' sold their stated speciality and supplied all sorts of men's apparel as well. Millinery warehouses, like the one on Jermyn Street, sold caps, aprons, and other similar accessories, advertised to be 'of the newest fashion ready made up, or made to order'.[115] Moreover, smaller London shops offered ready-made clothing and accessories as well, their goods commonly being supplied by larger warehouses. Mr Clowes of Conduit Street, for example, announced that in his shop 'is always kept the greatest variety of every different sort of cloaks, and all kinds of millinery, ready made; with a proper assortment of his patent interlined cloaks so superior for warmth and lightness. Good black mode cloaks, lined, wadded, and trimmed with neat lace or mock sable, 24s.'[116] The numbers of salesmen who operated in London also moved untold quantities of new and used clothing through the market, while many more small and medium tradesmen provided articles of clothing ready-made.

Several 'Wholesale Cloathes Warehouses' appeared as another component of the clothes trade in the 1790s.[117] As well, the occasional large lots of clothing would be auctioned to local tradesmen, and advertised both in London and the neighbouring districts in Kent. The size of the advertised consignments of clothing confirm, if confirmation were needed, the extent of the hitherto undetected trade in ready-made clothes. Two bales of 320 cloaks, jackets, drawers, and coats were included in one auction, in addition to 120 pieces of fabrics from which the clothing was probably made. The textiles were comprised of several sorts of heavyweight cottons, such as pillow, thickset, fustian, and velveret.[118] Another auction of the same sort announced that 200 lots of 'men, women and childrens' wearing-apparel were to be sold. The garments included in the sale were 'coats, waistcoats, breeches, shirts, hats, neckcloths ... gowns; silk and other petticoats; stays, shifts, sheets, pillow-cases, table cloths and variety [sic] of other articles'.[119] These auctions were ideal opportunities for tradesmen to

[115] *The Gazetteer and New Daily Advertiser*, 8 Sept. 1779.
[116] Ibid. 27 Oct. 1778.
[117] *The Universal British Directory* (1790), i, 56, 98, 300, 340.
[118] *The Morning Chronicle*, 6 Oct. 1784.
[119] *The Gazetteer and New Daily Advertiser*, 4 Oct. 1779.

acquire stocks of ready-made clothing at a discount, adding to the supply of clothes in circulation.[120]

London was more densely supplied with ready-made clothing, more attuned to fashion, and stood as a market of unrivalled size when compared to all other areas of Britain. London was unique, however, only in its extraordinary capacity. Major ports, particularly in the south of England, were well supplied with clothing manufacturers. (See Plate 25, John Keet's trade-card.) Moreover, the sale of ready-made clothes was nation-wide, common among the urban and rural provincial towns as well. The 1785 inventory of a bankrupt milliner, Elizabeth Brown of Norwich, reiterates this point. The shop appears to have been geared for the middling class of customer, and among the extremely detailed listing of her remaining stock there was an abundance of ready-made articles. It is possible to compare the value of yard goods as compared to ready-made stock; however, this would not present an accurate representation of the relative structure of the merchandise, since her stock had been allowed to run down drastically prior to the inventory. Nevertheless, it is interesting to note the selection of goods with which this unremarkable millinery shop furnished its customers. Not surprisingly, Miss Brown had a wide assortment of headgear: lawn caps, muslin caps, hair caps, children's quilted caps, 'Drest Capps', boys' caps, scarlet hoods for cardinals, white hoods for cloaks, black caps, bonnets, and a flowered-lawn cap. In addition there were matching cap-and-apron sets, sets of ruffles, many sorts of handkerchiefs, gloves, sets of purses and worsted caps, children's frocks, stomachers, muffattees, cloaks, and a quilted petticoat.[121]

Nearly 200 retailers were found in the *Universal British Directory* scattered throughout the country in large towns and small, ports and land-locked cities, all trading in ready-made clothing. Ports around the coast numbered many slop-sellers among their traders. Slop-sellers also flourished in places like Wantage and Maidenhead, miles from the

---

[120] Ready-made clothing seemed to offer the answer to the problem of clothing the poor of Hertfordshire, according to a proposal by one lady of the county. Her initial recommendation centred on the use of Sunday schools as a local source of cheaply sewn garments. However the author of the plan acknowledged that, 'Another way of providing the above ... suits of clothes and apparently the most eligible, is by purchasing at the Slop Shops in London, where they are ready made ... of a much better and more durable quality than those before mentioned, with only a small addition in price. BM 1609/1131, *Instructions for Cutting Out Apparel for the Poor* ... (1789), 56.

[121] Pamela Clabburn, 'The Provincial Milliner's Shop in 1785', *Costume*, 11 (1977), 100–10.

sea, but where the trade was known. Men and women owned clothes-shops in Coventry, Rochdale, Blackburn, Woodstock, and Richmond, to name but a few locations.[122] This listing only hints at the breadth of a ubiquitous enterprise. There were many other tradesmen and women who sold clothing in combination with other retail operations such as draper, dealer in used clothes, or leather-dealer, examples of which can be found in Birmingham. Further intriguing evidence of the sale of clothes in that city comes from an advertising trade-bill, illustrating a street scene in Birmingham, the corner of Ann and Congreve streets. (See Plate 26, Allin's trade-bill.) Allin's 'Cheap Clothes & York Shoe Warehouse' is shown topped by a large Union Jack and boasting a very large inscription on the wall—*Multum in Parvo*—with an equally contemporary meaning; much in little time, or for little money. A clear impression of Allin's business could not have come from a perusal of the directories of the time, in which he is listed simply as 'Taylor and Salesman'. The bill stands as a potent visual example of the clothing trade through the chance survival of this illustrated advertisement. Allin's was a large corner building with several substantial bay windows in which were displayed fashion illustrations in the bottom panes of glass and examples of clothing above; samples were hung against the outside wall for the perusal of passers-by in keeping with the practice of the trade.[123] Records of domestic tradesmen during this era are far from complete and the documentation which remains is suggestive rather than authoritative. None the less, the existing accounts confirm the presence of a trading community whose business was popular fashion, whose merchandise was ready-made and used clothing, and whose preoccupation would have been in providing the reigning modes for their customers.

One would expect the large port of Bristol to have tradesmen of this sort, and indeed there were six whose business involved the sale of clothes. The county seat of Worcester contained both a salesman, saleswoman and a draper-salesman; the market town of Banbury could even boast of a tailor-salesman in residence, while in Wirksworth Joseph Lamonby owned a shop he specifically designated as a 'Clothes Shop'.[124] The Morgan draper's ledger offers yet another indication of

[122] *The Universal British Directory*, ii, 620; iv, 298, 361, 673, 824, 947–9, 951; v, 20, 114, 218.
[123] John Johnson Collection, Men's Clothes, Box 1, Bodleian Library.
[124] *The Universal British Directory*, ii, 132–84, 208–41, 254; iii, 50, 799, 859, 863, 789–858.

the prevalence of ready-made clothing beyond the areas of the nation supplied by provincial centres or in close proximity to the capital. Among the many and varied items sold by the Morgans were all of the ready-made articles noted earlier: hats, gloves, mittens, velvet collars, shawls, both plain and muslin, stuff gowns, waistcoat shapes, waistcoats, quilted petticoats, cotton caps, shirts, pockets, breeches, cotton gowns, and chintz gowns.[125] The cost of the different sorts of gowns, from 5 shillings to 22 shillings, falls within the same price-range as gowns sold in other parts of Britain.[126] The appearance of this list of goods identical to those in urban centres, affirms that ready-made garments, produced as a by-product of the cotton industry (as well as other textile trades),[127] were sold nationally in response to the pressure for inexpensive, standardized fashions. All areas of Britain were supplied with cotton manufactures. The stimulus of popular fashion built up a demand for ready-made clothing that was satisfied with either local or regionally manufactured clothing. Trade like that of west-country clothes manufacturer John Totterdell (see footnote below), commands attention because of the size of the orders he was willing to undertake. But the foremost feature of Totterdell's enterprise is the insight it provides into the vast, ready-made trade in apparel associated with yet another of the textile industries in Britain. Homogeneity of demand, plus the standardization of tastes and popular fashion, opened all of Britain as a potential market for ready-made products.

THE MASS MARKET, POPULAR FASHIONS, AND CONSUMERISM

It is unfortunate that in studies which focus on the rise of the mass market in Britain—the appearance of this market, its characteristics, and the manner in which it was served—there has been an almost exclusive preoccupation with nineteenth-century developments. The

[125] Morgan Draper's Ledger, D/D ma 139, 7–11, 15, 41, 46, 50, 60, 64, 66, 68, 86, 87. Glamorgan Record Office, Cardiff.
[126] Ibid. 8, 11, 56, 60, 66, 68.
[127] West-country wool-draper, tailor, and merchant, John Totterdell of Bath, not only stocked ready-made suits and other garments, but also undertook to manufacture huge quantities of clothing. It appears that he routinely provided cargoes for export of this sort. But the probability of a home trade can also be inferred from Totterdell's remarks in an advertisement printed in 1766: 'Any gentlemen or Merchants, that chuses to contract for Five or Six Thousand Pounds worth of Cloaths, for Exportation etc., may have it made to Order, in four or five Months Time, by their most humble servant John Totterdell,' *Pope's Bath Chronicle*, 25 Sept. 1766.

assertions of historians of the pre-industrial period have been too often ignored. Joan Thirsk notes that 'It has become a convention to treat the mass market for consumer goods as a product of the Industrial Revolution, insignificant before the later eighteenth century'.[128] In some instances little importance is granted the mass market before the second half of the nineteenth century. The exploits of W. H. Smith, followed by the later successes of the provision and clothing multiples during the last quarter of the century, are presented as the archetypes of mass marketing by W. Hamish Fraser. He identifies two elements as crucial to the success of those late-nineteenth-century ventures: the recognition by traders of the profit to be made in serving a mass market, and a development in technology which permits mass production of inexpensive, quality merchandise. In the examples selected he cites technological breakthroughs in refrigerated shipping or mechanical sewing in the manufacture of clothing and shoes.[129] The implication is that these innovations in marketing and productivity were without precedent prior to the mid-nineteenth century. Fraser's study of the mass market looks too narrowly at the catalogue of items brought to the public, and does not assess the common elements of these ventures and the pivotal innovations over the preceding centuries. Are factory-made shoes, Irish butter, and Argentine beef more essentially consumer products of the mass market than stockings, gowns, corduroy breeches, chintz curtains, or muslin caps? Who would dispute that the volume of consumer items expanded during the later nineteenth century as problems of production and distribution were solved? But were the nineteenth-century advances in consumer production without precedent? One cannot look to the eighteenth century and expect to find the cumulative effects of an industrialization and mechanization characteristic of the next hundred years. Nevertheless, the eighteenth century witnessed the birth of entirely new elements in production and marketing: an unprecedented popular demand was founded on a popular appetite for new sorts of consumer goods apparent first in the late sixteenth and seventeenth centuries. The accomplishments of the eighteenth-century entrepreneurs in serving this first mass market cannot be discounted. The qualitative structure of the mass market was established prior to the Industrial Revolution. To claim the second half of the nineteenth century as the

[128] Thirsk, *Economic Policy and Projects*, 125.
[129] W. Hamish Fraser, *The Coming of the Mass Market, 1850–1914* (1981), chapters 8 and 9.

period of genesis of the mass market is short-sighted; observing the heavy flowering of that era, it ignores the roots and branches from which that prosperity sprung.

Firms of multiple shops, specializing in groceries, clothing, or shoes were built on the foundations of warehousemen like Taylor, or Allin's Birmingham clothes-shop. The creativity of marketing in the eighteenth century, with the proliferation of fixed shops, advertising, and display, confounds the assertion that prior to the mid-nineteenth century,

there was [not] much effort to attract customers, and the only sales techniques applied to the better-off shops was servility and deference. From about the 1870s onwards, however, there were signs of change, and of an awareness of new, potentially lucrative markets to be tapped. A few shopkeepers began to perceive that there was a place for cheapness and reasonable quality.[130]

Cheapness, fashionability, and 'reasonable quality' were the foundations of the cotton industry and were indeed catalysts in the establishment and enlargement of thousands of retail outlets. Street-sellers as described in Mayhew were not the sum of the popular retail establishments by 1850, as Fraser would suppose.[131] Consumers did not need to wait until the late nineteenth century to find stocks of new clothes, 'rather than revitalizing old'.[132] A century earlier the ready-made clothes trades flourished as an auxiliary of the cotton industry, as an aspect of retail trade in popular fashions. The mass market was well established by 1800, indivisible from industries which sustained consumer demand for popular fashions.

The cotton industry was one of the first to provide articles bought as a result of new sorts of needs, as from generation to generation the women and men of Britain set rising standards for their personal comfort and personal display. The mechanized cotton industry flourished in the provision of inexpensive, novel, and utilitarian textiles for this market. The success of the cotton industry in Britain would be replicated by many more producers with many other sorts of

---

[130] Fraser, 110.

[131] Gertrude Himmelfarb, *The Idea of Poverty*, 346–56, summarizes the misapprehension with regard to Mayhew's work. The title itself, *London Labour and the London Poor*, implies that he is studying a representative sampling of the working poor in the metropolis, where in fact he highlights exclusively the street people, the most peripheral elements of the working classes. To claim the retailing standards of this segment of London's labouring classes as representative of all suppliers of working people is unwise and inaccurate. To hold it up as a ruler against which all future developments could be measured results in qualitative distortions.

[132] Fraser, 101.

commodities. By the close of the eighteenth century, popular fashions of all descriptions had become a fixture of British economic and social life. The visual distinctions between the ranks narrowed as a result, blurring further social differentiation. The exigencies of fashion were now woven into the fibre of British society, and were institutionalized in periodicals, the structure of industries great and small, and the prevailing patterns of trade. Fashion and consumerism were inextricably linked to the future prosperity of the nation.

APPENDIX 1

Lancashire Textiles, 1775–1785

VARIETIES UNDER 12 PENCE

Checks
1 cotton-linen
8 cotton

Plain Cottons
9 sheeting
1 tick
1 roles
2 cotton

Stripes
8 Cotton
1 holland

Handkerchiefs
8 check
1 mallabar
1 romall
1 cotton
  mallabar check

Plaids
2 turkey

Fustians
1 figured diaper
3 jeans
4 pillows
1 towelling
1 figured dimity
1 corded dimity

VARIETIES FROM 13 PENCE TO 2 SHILLINGS

Checks
17 cotton
4 linen
1 half-cotton
7 furniture

Fustians
7 erminett
2 nankeen
9 pillow
3 cotton velveret
1 corded tabby
1 corded dimity
2 jeanett
1 drabett
1 jean
3 cotton thickset
1 sattinet
3 drawboy

Stripes
6 furniture
12 assorted cotton

Plain Cottons
7 sheeting
5 holland
2 bleached
8 cotton

Plaids
20 assorted
  cotton

Handkerchiefs
2 silk-muslin
1 silk-cotton

VARIETIES FROM 2 SHILLINGS TO 11 SHILLINGS

Cottons
1 plain
2 Marseilles quilting
1 flowered in the loom

Fustians
3 Genoa cord
1 King cord
1 Prince of Wales cord
2 Queen cord
2 corded tabby
1 Brunswick cord
3 cotton cord
1 cotton-faced cord
1 corded thickset
2 cotton rib
2 denim
1 Genoa velveteen
2 all-cotton velveret
1 spotted velveret
1 Genoa velveret
1 printed velveret
1 batwing velveret
1 printed jenet
2 barragon
1 sattinet

Stripes
1 cotton

Handkerchiefs
1 silk-cotton
1 silk-muslin
1 muslin
2 black yard
2 black
1 Barcelona

Checks
1 cotton

Sources: MC: MS ff 657 D43, Day Book of sales of an unknown Manchester firm, Manchester Public Library; Eng. MSS 1192, John Rylands Library, Manchester; *The Gazetteer and New Daily Advertiser*, 2 Aug. 1777; *The Beekman Mercantile Papers*, ed. Philip L. White (New York Historical Society, 1956), vol. iii.

APPENDIX 2

Ownership of Clothing from Contemporary Court Records

LIST OF COMPLAINANTS

1. May 1732. Benjamin Cook, seaman.
2. April 1743. Mary Stinson, kept house in Marsh Yard, Wapping.
3. June/July 1744. Thomas Meriton, captain of a vessel.
4. April 1745. Thomas Fox, ironmonger from Wolverhampton.
5. June 1745. Eleanor Read, lodgings-keeper.
6. October 1749. Thomas Leach, leather-seller.
7. October 1749. John Brown, coal-heaver.
8. April 1758. Mary Owen, servant.
9. June 1758. David Macquire, milkman.
10. September 1758. Abraham Baker, livery-stable operator.
11. September 1758. John Bagshaw, ostler.
12. January 1762. Mr Spencer, prosperous merchant(?).
13. September 1762. Mr Nichols, cider-merchant.
14. December 1762. Rachael Nanson, middling rank(?).
15. February 1763. Joseph James, merchant(?).
16. March 1763. George Hulbert, hair-cutter.
17. May 1763. Robert Pattison, pawnbroker.
18. July 1763. William Maclell, mariner.
19. October 1763. Mary Crookshank,(?).
20. November 1763. Patt Kirkeon, dealier in rabbits.
21. December 1763. ——, house in Virginia Street.
22. December 1763. Joseph Hobbs, shoemaker.
23. January 1764. Mrs Cotes, gentlewoman(?).
24. February 1764. David Scott, house-carpenter and ship-joiner.
25. February 1764. John Lockhart, sailor on merchant ship.
26. May 1764. Watford Runnet, furrier.
27. May 1764. William Thomas, servant; James Eaton, servant; Anne Brown, servant; Anne Robinson, servant; Jane Mitchell, housekeeper(?).
28. May 1764. James Yeoman, works for a refiner (sugar?).
29. January 1765. Mr Garratt, apothecary.
30. March 1765. Mr Edward Griffiths, . . ?
31. April 1765. C.H., baker for Rt. Hon. Lord Viscount Spencer.

Appendices

32. November 1765. ——, coachman for Mr Barrington Buggins, Esq.
33. January 1766. ——, coachman for Mr John Andrews, Esq.
34. February 1766. Mr Hutchinson, mathematical-instruments maker.
35. September 1771. Thomas Lacey, coachman.
36. December 1772. Samuel Yardley, tobacconist.
37. December 1772. Joseph Flocks, carpenter.
38. March 1774. George Westgate, 'poor servant fellow'.
39. December 1775. John Hollwell, mariner.
40. December 1775. Gannet Attkinson, wife of Crown prosecutor.
41. February 1776. Catherine Walker, gentlewoman; Isabella Berkeley,(?); Margaret Gardiner, servant; Elizabeth Wilson, servant(?); Mary Martin, servant(?).
42. April 1776. Robert Reynolds, carpenter and undertaker.
43. July 1776. Christopher Bones, servant.
44. September 1776. Solomon Fell, merchant.
45. July 1778. Margaret Robinson, traveller of modest means from Scotland.
46. December 1780. Judith Burdeau, daughter of a weaver.
47. December 1780. Ann Woodford, assistant to a nurse at St Bartholomew's Hospital.
48. February 1781. Elizabeth Croudy, servant.
49. July 1784. Elizabeth Coleman, seamstress.
50. October 1784. William Bunkum, captain of the *Venus*.
51. October 1784. Thomas Reed, hairdresser.
52. December 1784. Samuel Satcher, publican.
53. December 1784. Bartholomew Williamson, sailor.
54. December 1784. John Norfolk, fishmonger.
55. February 1785. Mary Taylor, servant to a milk-woman.
56. February 1785. Thomas Thrale, coach-maker.
57. September 1785. George Whitmill, agricultural labourer.
58. September 1785. Nathaniel Oliver, dyer.
59. February 1790. John Varney, grocer.
60. February 1790. William Vickery, driver of the Dover dilligence.
61. February 1790. John Morley, merchant's clerk.
62. April 1790. Margaret Finch, wife of 'a private soldier in the eleventh regiment of foot'.
63. April 1790. Thomas English, lodging-house keeper(?).
64. May 1790. Elizabeth Woodley, servant.
65. May 1790. Richard Turner, publican; Mary Lawrence, servant in 'The Bowling-pin'.
66. May 1790. John Ward, St Sepulchre's watchman.
67. July 1790. Countess of Berghausen.
68. September 1790. John Prior, carver and gilder.
69. September 1790. John Wishart, artisan(?).
70. January 1792. Sarah Bearcroft, widow in modest comfort.

71. May 1792. Thomas Gibbons, publican.
72. September 1792. John Perrot, butler.
73. September 1792. William Grace, house-owner and kept lodgers.
74. October 1792. Lady Ann Lindsay.
74. October 1792. John Manning, locksmith.
76. October 1792. John Norwood, coachman.
77. October 1792. Thomas Cruden, doctor of physic.
78. December 1792. James Palmer, 'poor man'.
79. December 1792. Mary Blackmore, wealthy widow.
80. December 1792. Mary Richardson, servant (upper).
81. January 1793. Charles Silk, metalworker for carriage harnesses.
82. January 1793. John White, labourer; Mary White, washerwoman.

LISTS OF CLOTHING TAKEN

1. May 1732. Benjamin Cook, seaman; goods stolen from on-board ship while it was docked in London:
> ... a Feather Bed, 2 Pillows, a Quilt, a Coat, a Waistcoat, a Jacket, and 2 Cotton Shirts ...

2. April 1743. Mary Stinson, kept house in Marsh Yard, Wapping. Goods stolen 'out of a Wash-Tub, in a close Yard belonging to my House ... on 22nd February, I saw the Prisoner with this Gown upon her Back; I found the Prisoner in Whitechapel.'
> 1 Cotton Gown, val. 2s. and 6d. one Dimitty Petticoat, val. 18d. 1 Apron, val. 1s. 1 piece of Linen Cloth, val. 1s. 2 Pillow biers, val. 1s. and 2 napkins, val. 6d. the Goods of Mary Stinson and one Pair of Stockings, the Goods of John White [lodger].

3. June/July 1744. Thomas Meriton, captain of a vessel—unspecified. Robbed by a gang of children.
> seven dimity waistcoats, value 3l.[£] two calico petticoats, value 3l. three petticoats, value 20s. a pair of stockings, value 4s. a petticoat, value 2s. and two tea spoons, value 4s.

4. April 1745. Thomas Fox, ironmonger from Wolverhampton; his trunk was stolen while Fox was back in Wolverhampton. He routinely rented a room in 'The Bell Inn', Wood Street when he was in town, leaving suitable clothes permanently at the inn. Unfortunately his goods proved too much of a temptation to a former porter at 'The Bell', who had been fired two weeks previously for drunkenness. On Fox's return he found the trunk missing.
> a cloth coat, value 30s. two cloth waistcoats, value 30s. a velvet waistcoat, value 40s. a velvet pair of breeches, value 5s. a pair of shag breeches, value 20s. ten holland shirts, value 3l. a hair trunk, value 10s. a pair of steel buckles, value 6d. and a book of accounts, called a ledger, value 6d.

The steel buckles are a nice touch—fashion, but without extravagance, in contrast to the high-quality clothing.

5. July 1745. Eleanor Read, lodgings-keeper; no indication of a husband. She had stolen:

> a cotton gown, val. 10s. a pair of shoes, val. 9d. a checked apron, val. 6d. and a cotton handkerchief, val. 6d.

6. October 1749. Thomas Leach, leather-seller; lived on Snow Hill with his wife. The theft was discovered by their live-in servant.

> 'I looked about and miss'd a silver spoon, also a child's cotton gown, not quite finished; a child's colour'd frock, straw hat, a cottong gown of my wife's, an apron, a black calamanco quilted petticoat.'

7. October 1749. John Brown, coal-heaver; Brown's wife took in washing and ironing. They lived in a house in Queen Street, Wapping. A woman came in saying she would iron the clothing and took it off. At the time Brown was 'lying on my own chest in my house and Joshua Fox sitting in a chair', both enjoying some brew. The goods stolen were:

> Two cotton shirts, one linen shirt, two linen aprons, two linen shifts ruffles, one dowlas shift, one huckaback table cloth; 'one of the cotton shirts belong to Joshua Fox, the other to James Richardson, he lodges in my house', Brown related.

8. April 1758. Mary Owen, servant; Mary Owen was recently down from Shropshire and stayed in a house owned by an elderly woman who was also from Shropshire. Owen had recently lost her place and was out looking for another position; when she returned she found that misfortune had struck again—her things were gone, stolen by Elizabeth Griffiths, also from Shropshire.

> One cotton gown, value 5s. three aprons, two handkerchiefs, two quilted petticoats, one linen shift, and one pair of sleeves were stolen. Later at the trial she related that, 'I happened to meet her in the street about a week afterwards; she had my gown and one apron on at the time. At first she denied the gown to be mine; but at last she owned it was mine.'

9. June 1758. David Macquire, milkman; Judith Macquire ... worked outside at unknown occupation and together they lived in a single room in a house in Great Swan Alley. Friends and neighbours next door, a baker and his wife kept a shop there, raised the alarm and notified Judith Macquire. On her return she noted the loss of:

> One linen gown, value 10s. one cotton gown, value 10s. one callimanco petticoat, value 10s. three linen shirts, one silk handkerchief, one pair of stays, one linen petticoat, two linen shifts.

10. September 1758. Abraham Baker: 'I ... keep post-chaises and saddle-horses to let out' and live in 'Haunch of Venison Yard, Lower Brook-street', Baker reported. A chest was stolen by a man that Baker had given the 'liberty

to lie a night or two in my hay-loft, he being a poor fellow'. The recipient of his charity was subsequently discovered 'with one of my coats and two waistcoats on his back'. The inventory of goods lost included:

> One wooden chest, value 18*d*. one fustian frock, value 10*s*. one pair of fustian breeches, value 3*s*. two cloth coats, value 10*s*. one pair bucksin breeches, value 7*s*. four waistcoats, value 3*s*. one pair of silver buckles, value 5*s*. one linen shirt, value 1*s*.

11. September 1758. John Bagshaw, ostler; kept his box in a closet at a lodging-house for safe-keeping while he was out looking for a position. Once he found a place he entrusted the lodgings-keeper with it still. Two weeks later a lodger walked off with the items and was later found with Bagshaw's hat on his head; and so he made his appearance before 'justice Fielding'.

> One cloth coat, value 8*s*. one cloth waistcoat, value 4*s*. one hat, one linen shirt, two pairs of cotton stockings, one pair of thread stockings, one common prayer book, one cotton cap, one linen handkerchief had been owned by Bagshaw.

12. January 1762.* Mr Spencer, prosperous merchant(?); lived at 6 Pennington Street, St George in the East parish, and suffered a substantial loss following a burglary:

> a large Portmanteau, containing one white Callico Gown, one ditto worked with a set Sprig ditto, one Callico printed with a yellow Ground ditto, one Long Lawn printed with a purple set Sprig ditto, one blue and white Linen ditto, one red and white ditto with a running Sprig ditto, one quilted upper Petticoat, three white Dimity under ditto, one white ditto Bed Gown, one striped ditto, one Dimity India Waistcoat, three Pair of white middle Sort of Sheets of the Beds, two Pair marked 1, 2, one Pair two W. and M. worked in Islet Holes, a large Parcel of Childbed Linen of all Sorts, four Pillow Cases, three large Diaper Table cloths marked S. two Napkins ditto marked S. two Russia Towels, six Irish Cloth Aprons, three Lawn and Muslin ditto, three Shifts, a large Parcel of small Linen, three Pair of Stockings, one white Sattin Hat wrapped up in a Piece of fine Holland, two good Flannel Under Petticoats . . .

13. September 1762.* Mr Nichols, cider-merchant; goods stolen from the house in Somerset Street, Whitechapel consisted of:

> A Pompadour coloured Coat and Breeches, the Coat lined with white; a blue Sattin Waistcoat lapelled, lined with white, Buttons the same Colour; and a very light-coloured Thickset Frock, with a Pocket in the Bosom.

14. December 1762.* Rachael Nanson, woman of modest competence; resided in Turnham Green and lost out of her house:

> One red Cardinal, one black Sattin Capuchin, one black Sattin Hat, one black Crape Gown, one black and white Cotton Gown, one white and

red Linen Gown, one blue and white Linen, one white Gingham, one Purple and white Bed Gown, one Yard of black and white Cotton, three quilted Petticoats, two under Petticoats, three Shift Bodies, two ditto with Sleeves, six Pair of Shift Sleeves, seven white Aprons, six check'd Aprons, 16 Caps, 11 Handkerchiefs, three Pair of Ruffles, one Pair of Stays, one Pair of corded Dimity Pockets ... [and] diverse other Things ...

15. February 1763. Joseph James, merchant(?); of Page Walk, Bermondsey, in Surrey, was robbed of the following:

one Damask Silk Gown, one green Silk Sack, one new fine Cotton Gown with large Flowers, red, green, and blue, one new fine Linen Sack Diamond Pattern, three Linen and Cotton Gowns, one sprigged, the other flowered, one new black flowered Sattin Cardinal, two Yards and a Half of black flowered ditto, two broad Mechlin-laced Handkerchiefs, one single, the other double, one Pair of new double laced Ruffles, three new fine India Silk Handkerchiefs, one red with a white Border, the other blue and white Check, three ditto Muslin ditto, one pair of Dresden Ruffles, a Box of new Ribbons, and several pieces of new Lace, one Sable Tippet, several new Callico long Neckcloths, one white and one red Sarsnet Handkerchief, trimmed with Gimp, one Pair of Womens new Stays white Callico, one large superfine scarlet Cardinal new, with Arm Holes, one Pair of white Silk Shoes flowered with Gold, one new Pair of black Callimanco ditto, upward of forty Men and Womens Shirts and Shifts, some ruffled and some plain, a great many Womens Caps and Handerkchiefs of various Sorts, some laced some plain, several Pair of Silk, Thread, and Worsted Stockings, two dozen and upwards of Muslin and Holland Linen of all Sorts, a Man's new Morning Gown, one Side Ruffel [sic] the other a fine Plaid, a new green Velvet Cap, several Remnants of white and green, and one Brocaded Silk ... a superfine light Cloth Waistcoat, three Pair of Breeches, one Pair of Cloth and two Pair of black Stockings.

16. March 1763.* George Hulbert, hair-cutter; was another of the thousands of people whose dwellings were rifled by thieves for clothing and other goods. His advertised loss covered two such incidents:

one Linen Gown with red and black Flower, one Purple and White Bed-gown, three Check Aprons, one Shift Body, and some time ago two red Duffle Cardinals, one black Sattin Bonnet, one Bed-gown, and a Piece of Check for an Apron.

17. May 1763.* Robert Pattison, pawnbroker; slept through a break-in when thieves entered the bed-chamber where he and his wife lay in their premises on Salisbury Street, Bermondsey. On his awaking he discovered the goods lost from their room included:

one Pair of Silver Buckles plain, and one Pair of wrought, and out of his

Breeches Pocket a Quarter of a Guinea, two Shillings in Silver, and a Key, a Piece of Silver representing Queen Anne, a Cotton Gown, a Damask Waistcoat, and a Tobacco Box, the Top Mother of Pearl, the Bottom Tortoiseshell ... and sundry other Sorts of Womens Apparel.

18. July 1763.* William Maclell, mariner; lost from his room in Deptford: thirteen white Shirts, two Check ditto, six Gowns, four Shifts, a Nankeen Waistcoat, eight pieces of new Linen Cloth, and diverse other Goods ...

19. October 1763.* Mary Crookshank, spinster; a wage-labourer of unspecified occupation, she lived in a lodging-house and had the following goods stolen by a friend of the lodging-house mistress:

... one cotton gown, one cotton handkerchief, and one pair of worsted stockings.

20. November 1763.* Patt Kirkeon, dealer in rabbits; residing in Hugger Lane, Queenhithe, Patt lost a quantity of money and some clothes of unspecified materials to a fellow Irishman, Thomas Bonham, who was described as 'a very sickly Man, with an old made Coat Black, with an outside Coat, very shabby Blue'. Patt Kirkeon lost:

three Guineas and a Half, out of a Wallet, one pair of Buck Leather Breeches, with three 5s. and 3d. Pieces in the Fob, three Shirts and three Pair of Stockings; a dark coloured Great Coat, and a scarlet Waistcoat with Brass Metal Buttons ...

21. December 1763.* ——, merchant(?). Although there is no name or occupation listed with this inventory stolen goods, the goods themselves are suggestive of a prosperous merchant family. Note the place of cottons in the wardrobe and the presence of prohibited Indian textiles:

Five plain Holland Shirts, marked H.K. ten Home-made ditto, marked D. four Nankeen Waistcoats never washed, new; one ditto washed one red and white striped Gingham ditto, one Pair of black Silk Stocking Breeches, with brown Holland Lining, one Pair of ditto Stockings, 7 or 8 Neckcloths with different Numbers marked H.K. one Piece of Home made Cloth, about 20 Yards; one striped India quilted Jacket double breasted, ties with Strings; one yellow India Silk Damask Gown, with a yellow Persian Lining with a small Pit of Sattin at the Bottom of the Apron, being the End of the Piece, with Joinings at the Top; one blue and white striped Lutestring ditto, with white Stuff Lining; two Inches ... before, one yellow Lutestring ditto, round Cuff, and open before, lined with yellow Persian, one grey Silk ditto lined with white Stuff, a Slip of grey under the Apron, and the Inside of the Breast black Silk, one Crape ditto mended under the Arms, one clear Muslin Apron two Breadths run with a String at the Top, one ditto whole with a striped border, run with a String, one thick Muslin ditto, two Breadths, bound at Top, four new Shifts Irish Cloth, one blue Silk quilted

Petticoat lined with white Stuff, one pink Stuff ditto, quilted with single Diamonds, one black Callimanco ditto lined with grey Stuff, and small double Diamonds, one plain single Lawn Handkerchiefs, one laced joining at the Breast ditto, two plain Clear Muslin Handkerchiefs, three or four striped boardered ditto, five red and white checked India Cotton Handkerchiefs, new, six ditto in Wear, three blue and white light and one dark ditto, five India Silk Handkerchiefs of different Patterns and two Cloth Aprons.

22. December 1763. Joseph Hobbs, shoemaker; while out with his wife one evening Hobbs was robbed of money and clothes by his journeyman, who also boarded in his house. The clothing was the property of Hobb's brother-in-law. Missing were:

one cloth coat, value 10s. one cloth waistcoat, value 10s. two India dimity waistcoats, and one ruffle shirt, the property of Samuel Jackman, and 5l. in money . . . the money of Joseph Hobbs . . .

23. January 1764.* Mrs Cotes, gentlewoman(?); a very large collection of clothing was stolen from the waggon of Mr Thomas Dullison. Mr Dullison was conveying this trunk and its contents from Coventry: between Dunchurch and London the trunk was found to be missing. The advertisement offering a reward for the missing goods asked that information be sent to Mrs Coates. She lived in Charles Street, Berkeley Square, and given the elaborate sophistication of her wardrobe she was probably travelling to London for the Season:

[*Gowns*] a scarlet and white Tissue Negilee and Petticoat, with a Tucker dewed in the Sack; a purple and green and white striped Lutestring Negligee and Petticoat, trimmed with a Garland of the same Colours, and a laced Tucker sewed in the Sack; a blue and white striped Lutestring Night-gown and Petticoat, trim'd with a Garland of the same Colours; a green Lutestring Night-gown, and a white Copper-plate Linen Night-gown, with green Sleeve-knots; a blue and white Linen Night-gown; a red and white gingham Night-gown; a fine white Callico Night-gown; a fine white Callico Bed gown and Waistcoat, both quilted; [*Clothing Accessories*] two Dimothy [dimity] Petticoats; four Pair of Fustian Dimothy Pockets; eight Pair of superfine Cotton Stockings, two Pair ditto, coarser; two Pair of fine white Silk Stockings; twelve fine Holland Shifts, and four Pair of Holland Sleeves ready made; a fine sprig'd flounced Muslin Apron; a fine spotted Muslin ditto; a flowered Gauze Apron; a plain Muslin ditto; two plain clear Lawn Aprons; a striped Muslin ditto, and a Callico Muslin ditto; four red and yellow spotted Silk Handkerchiefs; four Cambrick ditto white; two coloured Lawn Handkerchiefs, one spotted with red and white, the other check'd; three Cambrick white ditto; four Caps of spotted Lawn; three Pair of double Ruffles, one Pair striped Muslin, the other two plain; a white Sattin quilted Petticoat, and a blue Sattin quilted Petticoat; a white Sateenet Cloak, trimmed with Blond; an alamode black double Handkerchief, double laced; a Dozen and half of white Kid Gloves; two Pair of light brown ditto, and four Pair of Kid Mittens lined and work'd with different Colours; a Pair of purple and white Silk Shoes, with white Sattin Heels and Straps, never wore; two Pair of black Everlasting Shoes, quite new; two Pair of white Leather, and a Pair of black that has been worn; a Pair of new Tabby Stays; [*Miscellaneous*] a small Deal Box, which contained a Garnet and Pearl Necklace strung in a Pattern, and a Pair of clump Ear-rings to them, a white Necklace with four Rows, and a Pair of Ear-rings to them; a blue and white Necklace and Ear-rings; and a Pair of round Gold Ear-rings; a small Pocket Glass; two Common-Prayer Books; two ditto on the Sacrament; an English Dictionary; a Bundle of Cloth, a Parcel of Threads and Tape, a Machine filled with Hair-Powder, three or four Gallipots with Pomatums, etc. two Packing-cloths, six Towels, six Muslin Tuckers, a Parcel of Gloves and Mittens, that have been wore; and a Holland Combing cloth.

24. February 1764. David Scott, 'house-carpenter and ship-joyner'; lived in a house he owned in St Catherine, most probably with his wife. The goods lost consisted of:

one cloth cardinal, value 10s. one cotton gown, value 10s. one linen shift, value 3s. one shirt, value 3s. two neckcloths, value 4s. one pair of silver buckles, value 12s. one pair of pumps, value 2s. one callimanco petticoat, value 20s. one pair of linen sleeves, three linen aprons, one pair of cotton stockings, and two linen caps.

25. February 1764. John Lockhart, sailor on a merchant ship; Lockhart described the circumstances surrounding the theft of his clothing. 'The prisoner and I both belonged to the Devonshire, a merchant ship; she lay at Stone-stairs, by Ratcliffe-cross. On the 7th of January, I lost two guines, two dollars, and all my cloaths out of my chest, on board. . . . our captain took . . . [the prisoner] up eight days after. He had one of my jackets, and my breeches on, when taken he owned he had sold all the other things, and shewed me the houses where he had sold them, in Rosemary-lane.'

. . . one woollen jacket, val. 2s. one cotton jacket, value 2s. one pair of silver shoe buckles, value 10s. two silk handkerchiefs, value 2s. two shirts, value 2s. one pair of breeches, value 2s. a pair of stockings and one hat, value 1s. two dollars, value 9s. and two guineas . . .

26. May 1764. Watford Runnet, furrier; lived in a house in Black Friars, as a lodger. Returning home from work at about nine o'clock, he 'found . . . [his] box broken open' and clothes missing.

. . . one woollen cloth coat, value 3l. one woollen cloth waistcoat, with gold lace, value 50s. one pair of stocking breeches, value 5s. three shirts, two pair of silk stockings, one velvet waistcoat, one silk and cotton waistcoat, one man's hat, one pair of silver knee buckles with stones, two muslin neckcloths, and four linen stocks . . .

27. May 1764. William Thomas, servant; James Eaton, servant; Anne Brown, servant; Anne Robinson, servant; Jane Mitchell, housekeeper(?); all worked in the house of one George Rice, Esq. One evening after dark, as the witness reported, 'our nurse came down stairs, and said, the wind blowed her candle out, and she supposed the windows were not shut down, for the wind came down stairs'. James Eaton investigated and discovered that boxes containing clothes, stored in the servants' rooms at the top of the house, had been opened and the goods taken.

. . . one pair of pistols, value 12s. the property of the said George; four woollen cloth coats, value 40s. five woollen waistcoats, value 20s. three linen waistcoats, value 15s. three pair of cloth breeches, the property of William Thomas; one woollen cloth coat, one woollen cloth waistcoat, one pair of leather breeches, one man's hat, the property of James Eaton . . . one black silk cloak, one linen apron, one silk handkerchief, three linen shifts, three shift sleeves, four gowns, the property of Anne Brown, spinster; one linen pocket, one worsted pocket, two shifts, two aprons, one pair of stays, a cotton handkerchief, a silk handkerchief, three other handkerchiefs, a sattin hat, a pair of ruffles, the property of Anne Robinson, spinster; one long lawn gown, two cotton gowns, one callimacoe petticoat, one silk petticoat, four dimity petticoats, twelve linen shifts, three pair of ruffles, three linen aprons, four linen caps, two linen handkerchiefs, one pair of silver shoe buckles, and other things and one guinea and a half in money, the property of Jane Mitchell, spinster . . .

28. May 1764. James Yeoman, 'works for a refiner' (sugar?); Yeoman's workplace was in Foster Lane and he lodged 'in a house of the widow Baker in Crown Court', where he lived with his wife. Prior to the theft the couple had owned, among other things:

. . . one cloth coat, value 20s. one satin waistcoat, value 5s. one pair of leather breeches, value 20s. two pairs of worsted stockings, value 2s. one hat, value 7s. one cardinal, value 5s. one crepe tail of a gown, value 1s.

29. January 1765.* Mr Garratt, apothecary; advertised the theft of articles taken from his house in Stoke Newington shortly after the New Year:

Cups and Saucers, one red and white figured Tea-pot with Red Dolphins, one new green [?] Cloth. 12 Neck-cloths marked W. G. number 1 to 12, one Bird's Eye Diaper Tablecloth marked R.W.G. one Pair of plain Womans Ruffles, one ditto short double Lawn, one ditto India sprigged, single, one ditto double Book Muslin, one fine Ghenting Apron, one ditto plain, one ditto Muslin, 12 Pocket Handkerchiefs, one pair of Silver Buckles, one large Scarlet Cardinal, one small Tablecloth seemed [sic] up the Middle, marked g.4, one Woman's old black Cloak, lined with blue and black, one single square Handkerchief, five Neck Handkerchiefs, three Pair of fine Holland Sleeves, one laced Book Muslin Cap, four Ghenting Mobs [caps], six Muslin and Ghenting Mobs, two plain Shirts marked J.B. six Knives and Forks mounted in Ivory, Chinese taste; and three Pair of white Cotton Stockings, marked G.

30. March 1765.* Mr Edward Griffiths(?); in spite of the absence of any stated occupation the collection of items, including the silver, and the fashionable touches on many of the garments, suggests a level of prosperity that enabled a consumption of many of the niceties of the period. Mr Griffiths lived in a house in Gravel Lane, which was broken into and robbed of:

... six Ruffled Shirts, four marked E.G. one old plain ditto; two Cotton Gowns, grounded with black and white Flowers; two Cotton Slips, white ground and Purple Flowers; seven Pair of Shift Sleeves, seven white Aprons; one coloured ditto, three white Frocks; two Sheets, one marked with E.M.G. the other G. four Pocked Handkerchiefs, one red, spotted white; three white ditto, yellow, brown, and red; three Shift Bodies, two Pair of white Cotton Stockings; three Long Lawn Caps, Sheer Lawn Borders; five white border'd Handkerchiefs; one ditto new, double, and not hemmed; Several Night-caps and Tuckers ... one boys shirt; one new Sawn-skin Surtout Coat with broad Velvet Collar ... two silver Salts and shovels; one Scarlet Cloak; one old Sattin Hat ...

31. April 1765.* C.H., baker for Rt. Hon. Lord Viscount Spencer; the theft took place in an outbuilding at an estate in Wimbleton, Surrey. Lord Spencer offered a reward of twenty pounds for the return of the goods stolen, which included:

... twenty Pounds in Cash, chiefly Half Crown Pieces, contained in a Leather Purse, six Pair of white Stockings, some Linen and some Cotton, one Pair of black Leather Pumps, six Shirts nearly new, marked C.H. on the Hip, one ruffled, seven Cravats marked C.H. a Pair of Crimson Velvet Breeches with Gold Knee Bands, one Pair of plain knit buff colours ditto, new seated, one old scarlet Waistcoat with different coloured Buttons, one scarlet Flannel Waistcoat with [?] ..., plain Silver Buttons, one blue Surtout Coat with blue Basket Buttons, one light coloured old Fustian Frock, two new Fustian Frocks of a Lead Colour, Buttons the same, three new Silk Handkerchiefs in one Piece, blue Ground with white Spots, one Piece of new Nankeen, one red fine Flannel waistcoat lapelled, red Buttons, one Silk Plad Waistcoat, Buttons the same, and Fustian Back, two small Pieces of Flannel, with Tape Strings, one Plate Stock Buckle, and one Gold Ring enabled.

32. November 1765.* ———, coachman to a gentleman; employed as a coachman to Barrington Buggins, Esq. with a house in Philpot Lane,

Fenchurch Street, London and another premises in Wickenham, Kent. The coachman's room was rifled and the following goods lost:

...a black Surtout broad Cloth Coat Buttons the same; a black Silk waistcoat...a Pair of new solled [sic] Pumps, an odd Pump and two Pairs of other Shoes; a Pair of...Lamb Breeches very dirty; a coarse Fustian Frock with Ribbands to button at the Wrists; a new Fustian Frock and Breeches, and a Nankeen Waistcoat with Sleeves and a new Hat.

33. January 1766.* ——, coachman to a gentleman; while living in the coach-house of John Andrews, Esq., in Newman Street, Oxford Road, a robbery occurred. Stolen were:

...four Irish Shirts marked S.T. numbered 2, 4, 5, 6, two Neckcloth, one Marked T.M. one fine Callico Shirt, the Ruffles half off; one scarlet Cloth Cardinal, one black Silk figured ditto, two Cotton Gowns, one a white Ground, with running Sprigs and Flowers, the other a dark Purple, with white Strawberries in large Diamonds, one white Silk Hat lined with blue and blue Ribbons.

34. February 1766.* Mr Hutchinson, mathematical-instruments maker; his house was broken into and the following items stolen from his family:

...one striped Lutestring Gown, one Chintz Pattern Linen ditto without any Cuffs, and one new Silk and Cotton Handkerchief, not trimmed.

35. September 1771. Thomas Lacey, 'coachman to one Mr Broughton' who owned a livery stable; Lacey lived in Harford Street, Rag Fair and worked in the livery stable where the accused thief was also employed. Lacey recounted that:

'I lost a blue coat, ... it was my own; the waistcoat was the same colour. I lost a linen waistcoat besides, and a shirt, five neckclothes and a pair of leather shoes; two handkerchiefs and my razors, and a gilt metal buckle.'

36. December 1772. Samuel Yardley, tobacconist; Sarah Jordan was indicted for stealing the following items from his house:

...a silk gown, value 10s. two black silk cloaks, value 20s. a silk and stuff gown, value 5s. a white long lawn gown, value 5s. a white cotton gown, value 5s. a purple and white linen gown, value 10s. a sattin petticoat, value 10s. a dimity petticoat, value 2s. four linen napkins, value 2s. one pair of women's stays, value 10s. one linen shirt, value 1s. three linen aprons, value 4s, a muslin handkerchief, value 1s. and two black cloth waistcoats, value 2s. ...

37. December 1772. Joseph Flocks, carpenter; living in Brick Lane, Spitalfields. Flocks employed Sarah Wade as a servant, who was later convicted of the theft of the following goods from her master:

...a silver watch, a blue crape gown, value 15s. a linen gown, value 6s. a pair of jumps, value 4s. two linen handkerchiefs, value 2s. and a check apron, value 1s. 6d ...

Appendices 215

38. March 1774. George Westgate, 'poor servant fellow'; thus he described himself at the trial where he accused another servant with entering his room in 'The Bull' and stealing the following items:

> ... a red and white silk handkerchief, value 1s. and a muslin neckcloth, value 1s. ...

39. December 1775. John Hollwell, mariner; Hollwell came from Devonshire and worked on a ship that had docked at London. While he was employed at this work a fellow sailor went below and broke open Hollwell's chest, stealing:

> ... two check shirts, value six shillings, one linen shirt, value three shillings, one pair of mens leather shoes, value four shillings, one silk handkerchief, value three shillings, one pair of cloth breeches, value ten shillings, and one canvas bag, value six pence ...

40. December 1775. Gannett Attkinson, wife of the Crown prosecutor; even such as these were not proof against the thief. The Attkinson's house was broken into and they lost:

> ... a black negligee, blue paduasoy night gown, a pompadour silk gown, white lutestring gown and coat, a white muslin night-gown, two sets of ear-rings ... a cotton counterpane ...

41. February 1776. Catherine Walker, gentlewoman(?); Isabella Berkeley,(?); Margaret Gardiner, servant; Elizabeth Wilson, servant(?); Mary Martin, servant(?); Mrs Walker lived at Felton Hill and had a box of clothing belonging to herself and her servants sent via the Chertsey wagon 'the Earl of Loudoun's Whitehall' to Margaret Gardiner, who awaited it there. The wagon stopped at 'The New White Horse Cellar' in Picadilly where the box was to be unloaded and stored. It was at this place the trunk was stolen. The contents included:

> ... a spotted muslin sacque and petticoat, value twelve shillings, a pink satin quilted petticoat, value three shillings, a pair of womens stuff shoes, value one shilling, and a pair of silver shoe buckles, value ten shillings, the property of Isabella Berkeley, spinster (?); a scarlet and white sattin gown trimmed with black lace, and a white lace tucker, value 7l. the property of Margaret Gardiner, spinster; a Manchester quilted bed gown, value four shillings, two linen night caps laced, value five shillings, a pair of white sattin shoes, value five shillings, twenty-five silk handkerchiefs, value thirty shillings, a pair of treble lace ruffles trimmed with valencienne, value forty shillings, a pair of double point lace ruffles, value forty shillings, two pair of double worked muslin ruffles, value eighteen shillings, three lace tuckers, value twenty shillings, two pair of short double ruffles, value ten shillings, four yards and a half of new lace, value eighteen shillings, a net double handkerchief, value twenty shillings, a striped muslin apron, value four shillings, two yards and half of pattinet [sic] gauze, value seven shillings, a hair trunk, value five shillings, and a cotton night waistcoat, value three shillings, the property of Catherine Walker, spinster; a dark coloured

cotton gown, value fourteen shillings, a worked muslin apron, value two shillings, a linen apron, value two shillings, and two pair of linen shift sleeves, value two pence, the property of Elizabeth Wilson, spinster; a black stuff petticoat, value two shillings, the property of Mary Martin, spinster . . .

42. April 1776. Robert Reynolds, carpenter and undertaker; Reynolds lived in Chiswell Street, Moorfields in a house he owned. Coming home about eight o'clock he encountered two men coming out of his house, a house he had left double-locked. One man held a candle and the other a bundle in one hand and a leg of pork in the other. The first man pushed the candle into Reynolds [face] when he spotted him and both thieves ran off. Luckily the man carrying the bundle dropped it and was ultimately caught. The goods taken consisted of cash and clothing owned by Reynolds:

. . . 12 linen shirts 24s; a silver watch capped and jewelled £8; seven silver tea spoons 10s; two silver table spoons 10s.; a silver milk pot 10s.; two pair of silver shoe buckles 12s.; two pair of silver studs 2s.; a pair of stone ear rings, set in silver 2s.; eighteen linen handkerchiefs, 18s.; six silk handkerchiefs 12s.; three linen shifts 12s.; a black silk cloak trimmed with black lace 6s.; a blue silk cloak trimmed with white lace, 6s.; a silk cloak 4s.; a woman's linen waistcoat 5s.; a linen bed gown 2s.; a gauze apron 2s.; a pair of laced linen ruffles, 4s.; three children's linen gowns 3s.; a linen tablecloth 2s.; a linen napkin, 1s.; a woman's linen shirt 2s.; a cheque linen apron 1s.; sixteen yards of Irish linen 24s.; a long lawn gown 3s.; a slated leg of pork 25s. . . .

43. July 1776. Christopher Bones, servant; Bones was accosted by an out-of-work servant who asked him if he knew where there was work. Falling into conversation, the two men drank cider together in Islington. Bones and his new-found friend then went to the Swan Yard in the Strand where a friend lived and at this point, as Bones relates the story, 'I sat down on the stairs, and having been drinking in the morning without eating, I was very sick and fell asleep'. Upon awakening he discovered the bundle with his clothes missing along with his drinking partner. Bones went in hot pursuit and soon discovered that his quarry had left London on the Northampton fly [coach]. Bones persevered in his hunt: 'a man lent me a horse, and I overtook it [the coach] at Islington; the prisoner was on top of the coach, and had my neckcloth, shirt and waistcoat on . . . he confessed he had pawned the other things.'

. . . one linen shirt, value 5s. one muslin neckcloth, value 1s. one nankeen waistcoat, value 2s. two linen handkerchiefs, value 1s. two pair of silk stockings, value 5s. one pair of thread stockings, value 2d. one pair of silver shoe buckles, value 10s. and one pair of silver knee buckles, value 2s. . . .

44. September 1776. Solomon Fell, merchant; found his city house robbed while he was away in Norfolk. In addition to silver, the clothing lost was listed as:

... four woollen cloth coats, value 8*l*. four woollen cloth waistcoats, value 3*l*. four pair of woollen cloth breeches, value 40*s*. a man's cloth night-gown, value 20*s*. two cotton waistcoats, value 10*s*. twenty-four linen shirts, value 24*l*. twenty-four muslin stocks, value 40*s*. six morning caps, value 6*s*. twenty-four pair of silk stockings, value 9*l*. twelve linen handkerchiefs, value 30*s*. a man's hat, value 15*s*. two pair of leather shoes, value 17*s*. . . .

45. July 1778. Margaret Robinson,(?); travelled from Leith, Scotland on a coastal vessel down to London. Her bundle of clothing was stolen by a fellow passenger. The bundle contained:

... a stuff petticoat, value 6*s*. a linen shift, value 4*s*. a laced cap, value 3*s*. a pair of stone sleeve-buttons, set in silver, value 3*s*. a pair of plated shoe-buckles, value 2*s*. two cotton handkerchiefs, value 5*s*. a silk handkerchief, value 4*s*. and a pair of scissors with silver handles, value 2*s*. the property of Margaret Robinson, spinster.

46. December 1780. Judith Burdeau, daughter of a weaver, Claude Burdeau. The Burdeaus lived in Hare Court, Hare Street, Bethnall Green. Judith noticed her clothes on the shelf in the closet before she went to work up in the 'one pair of stairs room'. 'I buttoned the door after me and went to work up one pair of stairs higher'. The theft occurred some hours later. Her clothes were taken out of a closet below the upper floor where she worked, but the thieves were stopped by a neighbour. They were charged with the theft of:

... two cotton gowns, value 38*s*. a silk gown, value 6*s*. a cotton gown, value 1*s*. 6*d*. a pair of stays, value 8*s*. a stuff petticoat, value 7*s*. a pair of cotton stockings, value 6*d*. . . .

47. December 1780. Ann Woodford, 'assistant in the day time to the nurse of one of the wards of St. Bartholomew's Hospital'; the night nurse was accused by Woodford of pilfering from her locker and removing some clothing. 'The prisoner was there as a night nurse in the beginning of July. About five in the morning she went round the ward and asked the people if they wanted any thing; she found none awake but me. I asked her what it was o'clock, she said it was not five. I lay down again and went to sleep; I waked about half past five and missed the prisoner out of the ward; I got up and went to a locker to look for the things I had pulled off after dinner on Sunday; I found it empty'. Woodford had lost:

two cotton gowns, value 1*l*. 2*s*. a white lawn apron, value 1*s*. 6*d*. a check apron, value 1*s*. 8*d*. a linen handkerchief, value 1*s*. 6*d*. and a linen cap, value 6*d*.

48. February 1781. Elizabeth Croudy, servant; who in the course of her move from Black Swan Court, opposite St Paul's, to Mr Holt's in Chancery Lane, had her box stolen. As she was knocking on the door of her new place of employment, someone made off with her box, which contained:

... seven yards of cotton for a gown, value 19s. eight yards of linen cloth, value 5s. two stuff gowns, value 15s. two cotton gowns, value 10s. two linen gowns value 40s. a silk gown, value 40s. a crape gown, value 10s. two pair of women's stays, value 20s. a Marseilles petticoat, value 10s. three dimity petticoats, value 10s. three stuff petticoats, value 14s. six linen shifts, value 15s. ten linen aprons, value 10s. a muslin apron edged with lace, value 15s. a muslin handkerchief edged with lace, value 3s. a worked muslin handkerchief, value 5s. six muslin handkerchiefs, value 18s. two silk handkerchiefs, value 8s. four linen handkerchiefs, value 5s. two pair of linen mitts, value 5s. a pair of leather gloves, value 2s. two paper boxes, value 6d. a silk cloak, value 20s. three pair of silk stockings, value 6s. and three pair of cotton stockings, value 4s. the property of Elizabeth Croudy, spinster ...

49. September 1784. Elizabeth Coleman, seamstress; lost a trunk and its contents during a fire, when the rescued trunk was spirited away by a thief during the excitement.

... one pair of women's stays, value 5s. one striped apron, value 8d. a paper fan, value 2d. an iron fork, value 1d. three gauze handkerchiefs, value 3d. and two pieces of printed linen, value 2d. one pin cushion, value one half penny, one needle book, value one halfpenny, one pair of stockings, value 6d. and one hair trunk, value 6d. ...

50. October 1784. William Bunkum, captain of the *Venus*; the steward was on board when the theft was discovered and he raised the alarm. The captain lost a considerable quantity of clothing:

... four dozen shirts, value 10*l*. six other shirts, value 12s. three coats, value 20s. eight dimity waistcoats, value 4*l*. four pair of nankeen breeches, value 20s. three pair of silk breeches, value 20s. three pair of silk stockings, value 20s. eighteen pair of cotton stockings, value 20s. one pair of boots, value 5s. six pair of thread stockings, value 6s. eighteen pair of cotton trowsers, value 30s. five silk handkerchiefs, value 15s. one linen stock, value 1s. four stocks, value 2s. six linen handkerchiefs, value 3s. one pair of sheets, value 5s. one table-cloth, value 8s. one silk umbrella, value 5s. one gold ring, value 10s. one silver stock-buckle, value 5s. one pair of stone knee-buckles set in silver, value 2s. the property of the said William Bunkum.

51. December 1784. Thomas Reed, hairdresser; in addition to the silver teaspoons, buckles, and similar items, Reed also lost clothing invaluable to his calling, particularly if his clientele were of the fashionable world:

... sixteen calico shirts, value 8*l*. sixteen linen stocks, value 8s. six linen handkerchiefs, value 12s. two sattin waistcoats, value 20s. four cloath coats, value 6*l*. two striped silk waistcoats value 20s. two dimity waistcoats, value 15s. two pair of black sattin breeches, value 20s. six pair of silk stockings, value 20s. ten pair of silk and cotton stockings, value 50s. three table cloths, value 20s. ...

Appendices

52. December 1784. Samuel Satcher, publican; Satcher kept 'The Black Dog' in Shoreditch, where he lived with his wife and mother. The family was robbed of silver spoons, china, and a snuffbox, in addition to personal clothing.
   . . . two silk cloaks, value 10s. three muslin aprons, value 10s. three Pair cotton Stockings, value 1s. three cotton handkerchiefs, value 2s. two Silk handkerchiefs, value 4s. one Pair gloves, value 1d. . . .

53. December 1784. Bartholomew Williamson, sailor; had his clothes stolen while he slept the sleep of the dead drunk.
   . . . one woollen jacket, value 3s. one Pair fustian breeches, value 3s. one handkerchief, value 12d. one Pair knee buckles, value 4s. one Silk purse, value 12d. one velveret waistcoat, 3s. one Pair cotton stockings, 12d. one Pair silver buckles 20s. one Pair silver sleeve buttons, 12d. . . .

54. December 1784. John Norfolk, fishmonger; lived in White Rose Court, Coleman Street, with his wife and a servant. The servant made off with:
   three gowns, value 20s. a silk petticoat, value 3s. one dimity petticoat, value 3s. five shirts, value 20s. four shifts, value 8s. three bonnets, value 3s.

55. February 1785. Mary Taylor, 'servant to a milk-woman'; had her box and its contents stolen from her.
   . . . one dark coloured cotton gown, value 14s. one cloth cloak, value 8s. four caps, value 4s. two pair of stockings, value 2s. three pair of worsted stockings, value 1s. one black silk handkerchief, value 1s. one muslin handkerchief, value 1s. one linen handkerchief, value 6d. one pair of shoes, value 6d. . . .

56. September 1785. Thomas Thrale, coach-maker; lived in Long Acre and had stolen from his house:
   . . . one silk gown, value 5s. a cotton gown, value 5s. a silk petticoat, value 2s. a dimity petticoat, value 2s. a marseilles petticoat, value 5s. two linen table cloths, value 6d. a marseilles waistcoat, value 5s. two muslin aprons, value 5s. a fringed shawl, value 6d. a marseilles waistcoat, value 2s. a pair of boots, value 2s. a pair of cotton stockings, value 12d. one shift, value 12d, [?] four small pieces of muslin, value 12d. . . .

57. September 1785. George Whitmill, agricultural labourer; while bringing in a crop other labourers noticed the activity of the thieves and gave chase through the fields. Whitmill was fortunate to recover the items he had lost even before he knew he had lost them.
   . . . one cotton gown, value 17s. a camblet gown, value 10s. a linen gown, value 2s. a cloak, value 5s. a pair of leather breeches, value 3s. a linen waistcoat, value 9d. a piece of linen, value 1d. a shift, value 2s.

58. September 1785. Nathaniel Oliver. dyer; lived in a small house with his wife and children. His house was robbed of:
   . . . one cloth apron, value 12d. one linen shift, value 12d. four children's

shifts, value 18*d*. one callico bedgown, value 9*d*. one diaper napkin, value 8*d*. ten children's caps, value 2*s*. one muslin cap, value 3*d*. one muslin handkerchief, value 6*d*. two callico handkerchiefs, value 6*d*. . . .

59. February 1790. John Varney, grocer; he and his wife found the servant, Jemima Wilson, gone with a quantity of their clothing. She was supposed to start on the washing early one morning but instead broke open the wash-house and fled with the clothes. Varney's daughter was left with nothing to put on following this catastrophe. The total loss came to £3, and Mrs Varney and her daughter enumerated some of the items taken during their testimony in court:

'I missed my own, a silk cloak and bonnet, and check apron; I saw them on the night before behind my chamber door, . . . I know the cloak by a bit of a darn on one side of it . . . I had it about two years; there is a callico petticoat, of my daughter's I can swear to; and a linen gown and coat, of her's . . .'. [Rebecca Varney:] 'I can swear to a callico gown, and the other two gowns, and my stays . . .'.

60. February 1790. William Vickery, driver of the Dover diligence; a bundle of his clothing that he had put on the seat of the coach was stolen by a light-fingered passer-by.

. . . one shirt, value 3*s*. a pair of cotton stockings, value 3*s*. a pair of leather shoes, value 2*s*. 6*d*. a muslin handkerchief, value 2*s*. a linen handkerchief, value 18*d*. a pair of leather boots, value 3*s*. 6*d*.

61. February 1790. John Morley, merchant's clerk; John and Ann Morley, his wife, and their child, lived in Princes's Row, Stepney, in a house that Morley owned. The loss was discovered by Ann Morley, who described the goods stolen during the trial:

'I looked in a closet by the fire-side, and I missed four tea-spoons, and a breakfast napkin; I went up stairs, and observed a large deal box that was behind the door, drawn out in the middle of the room, with a drawer upon it; and one muslinet gown and coat, and one callico lawn gown, with long sleeves [were missing]; I then opened my husband's chest, where his clothes were, and I perceived it empty; there were five cloth coats, a sattin waistcoat and breeches, a pair of black sattinet breeches, a pair of new blue cloth breeches, a pair of corderoy breeches, a pair of leather breeches, a pair of black cloth breeches, and several waistcoats.'

62. April 1790. Margaret Finch, wife of 'a private soldier in the eleventh regiment of foot'; her box was lost out of the taproom in 'The Queen's Head', Billingsgate. Margaret Finch explained the events in her testimony. 'I saw the prisoner take it; I was giving my child something to eat: the prisoner and one North said they would take me to Graves-end; but the boat was not fit; and they took my box; I got a pint of porter in a bottle, to take down with me; they never returned; we found the box broke open, at Rotherithe, about two miles off . . .'. Margaret Finch lost from her box:

... seven pair of cotton stockings, value 14s. two cotton gowns, value 20s.

63. April 1790. Thomas English, lodging-house keeper; 'I keep a house in Great Earl-street, Seven Dials. The prisoner lived with us three weeks, as a servant; and went away that night about seven. At ten o'clock I went up stairs, and found the back door open that goes on to the leads ... I called ... [my wife], and told her; and she said, then I dare say all my clothes are gone.' Mrs English was correct in her guess and lost:

> three cotton gowns, value 14s. four cotton petticoats, value 10s. two black silk cloaks, value 10s. and one childs dimity cloak, value 2s.

64. May 1790. Elizabeth Woodley, servant; this widow worked for Mr Gilding, 113 Aldersgate Street, when on the sixteenth of the month a fire broke out. Elizabeth Woodley's things were piled on the street where they attracted the attention of 20 year-old James Flindell who carried away the chest of drawers, only to be stopped by the constable of Clerkenwell. The clothing in the drawers was listed as:

> one yard and a half of muslin, value 5s. one silk gown, value 2l. one silk petticoat, value 18s. three cotton gowns, value 4l. 4s. a petticoat, value 10s. a gown, value 10s. a bed quilt, value 10s. 6d. a ditto, value 5s. a petticoat, value 4s. a shawl, value 3s. two yards of lace, value 5s. two yards of callico, value 4s. a bed-gown, value 3s. a handkerchief, value 1s. a pair of stockings, value 1s. a pair of pockets, value 1s. three linen towels, value 3d. a piece of cloth, value 8s. three other pieces of cloth, value 6s. ...

65. May 1790. Richard Turner, publican of 'The Bowling Pin'; Mary Lawrence, servant to publican; she worked in 'The Bowling Pin', in Bowling Pin Alley. During the night of 27 April a lodger broke into the building and ran off with a bundle of goods, some of which belonged to Richard Turner and his wife and the rest to the unfortunate servant.

> ... nine linen shirts, value 27s. five shifts, value 10s. eight pairs of cotton stockings, value 8s. six aprons, value 6s. a muslin shawl, value 1s. a linen gown, value 6s. his property; and one linen gown, value 6s. three silk handkerchiefs, value 3s. three muslin handkerchiefs, value 2s. one callico shawl, value 2s. three caps, value 2s. two shifts, value 4s., the property of Mary Lawrence ...

66. May 1790. John Ward, St Sepulchre's watchman; Ward was married with at least one child, and he and his wife lived in lodgings of two rooms. One room was broken into and the articles listed below were stolen:

> ... one man's woollen cloth coat, value 8s. and a boy's ditto, value 10s. one pair of nankeen breeches, value 4s. four linen waistcoats, value 5s. one dimity waistcoat, value 2s.

67. July 1790. The Rt. Hon. the countess of Berghausen; she was involved in a

very complex case wherein her common-law husband (describing himself as a major in the Irish Volunteers) attempted to steal several trunks of clothes and other goods, prior to running off with another woman whom he described as his cousin. These trunks had been packed for a trip to Norfolk, and one can assume would contain all the items deemed essential for a trip of one or two weeks:

> a silk gown and coat, value 40s. a brocade gown, value 40s. a green gold and silver silk gown and coat, value 42s. a court dress, value 42s. a scarlet ditto, value 42s. a black silk gown and coat, lined with lilac silk, value 42s. a sattin ditto, value 42s. a black sattin ditto, value 42s. a yellow ditto, value 42s. a flesh coloured ditto, value 42s. a rose coloured ditto, value 42s. a snuff coloured ditto, value 40s. a gold embroidered gown and coat, value 42s. a muslin ditto, value 30s. a white ditto, value 30s. a dimity petticoat, value 5s. one other worked petticoat, value 8s. two ditto, value 8s. two plain ditto, value 8s. a pair of white sattin shoes, value 2s. thirty-eight shifts, value 38s. a white gown, value 30s. eight petticoats, value 40s. two table cloths, value 5s. two rose coloured silk jackets, value 40s. one sattin gown and petticoat, value 40s. one ditto, value 40s. one yellow jacket and petticoat, value 40s. one sattin ditto, value 20s. one ermine petticoat, value 5l. one flowered ditto, [petticoat] value 40s. one lilac ditto, value 40s. one yellow ditto, value 5s. one white ditto, value 5s. one blue lutestring ditto, value 20s. one maroon ditto, value 20s. one black silk ditto, value 40s. a woman's gauze gown and petticoat, value 20s. a green silk riding dress, value 20s. a red silk ditto, value 5s. a green velvet riding dress, value 20s. an orange and scarlet ditto, value 20s. seven shifts, value 7s. twelve pair of silk stockings, value 12s. two black silk cloaks, value 20s. a worked muslin cloak, value 12s. five table cloths, value 5s. two gowns and petticoats, value 20s. two pair of shoes, value 5s. a star and order, set with brilliants, value 5l. a pair of pearl drops, value 50l. two rings, value 40s. a pair of silver candlesticks, value 5l. a silver tea-urn, value 20l. . . .

68. September 1790. John Prior, carver and gilder; during the process of moving from Shoe Lane to No. 4 Stephen Street, Rathbone Place, Prior left a box at 'The White Lion' in Shoe Lane. It contained the following:

> eighteen yards of black silk lace, a cotton gown, value 12s. five yards and a half of muslin, value 40s. a shirt, value 5s. a shift, value 4s. four muslin neck handkerchiefs, value 12s. a shawl, 4s. 6d. and two hundred steel needles, value 1s.

69. September 1790. John Wishart, artisan(?); Wishart is the hero of a cautionary tale whose story is best told in his own words. 'I lost the things in the indictment from the lodgings of the prisoner [Susannah Corbett]: I had been out with some friends: I went home with the prisoner between twelve and one at night: I think I met her at Cripplegate: I was not sober: I was locked

out: . . . the prisoner said, she had a room to herself: I went there; a young women brought in some liquor: I went to bed, and fell asleep, I awaked, and found myself alone, and my clothes gone . . . it was about two o'clock when I got out of the house, and got to the watch-house, in Bunhill-row . . .'. John Wishart's clothing consisted of:

> . . . one cloth coat, value 20s. one waistcoat, made of silk and cotton, value 5s. one pair of nankeen breeches, value 5s. and one silver stock buckle, value 5s. . . .

70. January 1792. Sarah Bearcroft, widow(?); Sarah and her three children lived in a house at No. 1 Helmet Court, in the Strand, owned by her brother, a bookbinder, who also lived there with his wife and children. Sarah's portmanteau had been left in the side parlour. She and her brother had gone out, and on returning, 'it was gone and the window belonging to the side parlour was lifted up'. The clothing she lost was listed as:

> . . . one dimity gown, value 5s. two dimity petticoats, value 3s. three linen frocks, value 3s. six linen frocks, value 1s. a silk petticoat, value 5s. two linen napkins, value 2s. four linen clouts, value 12d. four silk gowns, value 4l. four linen gowns, value 40s. four cotton gowns, value 40s. one silver watch, value 40s. a piece of baize, value 12d. and a linen shift, value 6d. . . .

71. May 1792. Thomas Gibbons, publican; Gibbons was a widower and lived with his sister and mother-in-law at 'The Folly House', Blackwall. It appears to have been a prosperous business. On Easter night at ten o'clock some men knocked on the door and asked for something to drink. After letting them in— licensing hours did not seem to be a concern— they attacked the landlord and 'they swore they would blow my brains out'. Gibbons, his mother-in-law, his sister, and a returning servant were all tied up and their heads covered. The thieves spent three or four hours 'plundering the bar' and the house: the mother-in-law, Jane Mole, complained that in addition, 'after they had taken what they pleased, they supped upon cold roast pork, and sucked 15 eggs'. The articles stolen included clothing and other valuables:

> . . . a silver watch, value 30s. one pair of shoes, value 2s. two silver table spoons, value 16s. nine cotton gowns, value 9l. two gold wire ear-rings, value 2s. a pair of shoe and knee buckles, value 5s. and three guineas in monies . . . the property of Thomas Gibbons; seven gowns, value 3l. three black cloaks, value 2l. one red ditto, value 10s. nine silk handkerchiefs, value 18s. two tuckers, value 2s. one half guinea, and a Spanish dollar, value 4s. 6d. the property of Eliz. Gibbons; two silk gowns, value 20s. three cotton gowns, value 30s. a silver watch, value 30s. three linen shifts, value 10s. two black silk cloaks, value 20s. six guineas in monies . . . and a bank note, value 10l. the property of Jane Mole.

72. September 1792. John Perrot, 'butler to Mr. Bushnell'; John Perrot'

clothing was stolen from the butler's pantry during an hour of the day when Perrot was busy. Fortunately the culprit was spotted on John Street by a 'gentleman's coachman' who raised the alarm. The list of goods lost was as follows:

> ... a cloth coat, value 2s. four cotton waistcoat, value 2s. a pair of leather shoes, value 1d. the goods of John Perrot, and a linen table cloth, value 2s. the goods of John Askeel Bushnell, Esq.

73. September 1792. William Grace, house-owner and kept lodgers; Grace lived with his wife in Black Horse Yard, East Smithfield. Grace's wife was busy with a sick woman lodger and he went to bed, hanging his clothes on a line in his room. The next morning he found the bolt on the door forced and the clothing missing.

> ... a cloth coat, value 20s. a clothe [sic] waistcoat, value 10s. a pair of velveret breeches, value 5s. a stuff gown, value 2s. a man's hat, value 2s. a man's wig, value 5s. a muslin neckcloth, value 1s. a linen handkerchief, value 1s. a cottong bed-gown, value 1s. a pair of man's shoes, value 2s. a pair of plated buckles, value 2d. a linen apron, value 1s. a linen waistcoat, value 1s. a velveret hatband, value 2d. a brass hat buckle, value 2d.

74. October 1792. Lady Ann Lindsay; one of her servants stole several pieces of yard goods that Lady Ann had bought to make up into clothes. At the trial Lady Ann testified that, 'these articles were in a parlour, in order to be made up; I missed some on the 10th of October, and others at various times.'

> ... two yards and a half of printed cotton cloth, value 5s. four yards of printed cotton for bordering, value 4s. and four chair covers ...

75. October 1792. John Manning, locksmith; Manning lived in Little White Lion Street, Seven Dials, and worked at a shop in No. 12 Castle Street. In the same building was a woman who did laundry. Manning discovered that he had lost two items of clothing stolen by the laundress.

> nankeen breeches and a muslin neckcloth.

76. October 1792. John Norwood, 'coachman to Mrs. Gibson in Hertford-street, Mayfair'; Norwood left some of his clothing in a box at 'The Chesterfield Arms' and later discovered his loss when he went to check on his property.

> ... a pair of velveteen breeches, value 5s. two linen shirts, value 14s. and two dimity waistcoats, value 4s.

77. October 1792. Thomas Cruden, doctor of physic; Cruden lived at Ten Court, Fenchurch Street, and had a bundle cut from the back of his phaeton [a sporting carriage] while it travelled through the streets.

> ... a cotton quilted morning gown, value 20s. a flannel gown, value 20s. and another cotton morning gown, value 10s. and a linen towel ...

78. December 1792. James Palmer, described by the court as a poor man;

Palmer lived in one room in a lodging-house in Short's Gardens, Drury Lane, with his wife and two children. Palmer's wife had been ill and they had engaged Susanna Edwards to care for her. She 'lived in the same room likewise with us'. Palmer testified, 'I went to be at eleven o'clock, and fell asleep; I had not been in bed above three or four hours, before my wife alarmed me, and asked where Sukey was, as she could not feel her by her side, or hear her breathe.' Susanna had slipped out with a bundle of goods and money taken from Palmer and his family. The prisoner's defence at her trial was that she had been given the clothing by Mrs Palmer since they had no money to pay her. Mrs Palmer, however, replied that, 'with respect to giving her my shift, or my shoes, then I must have gone without myself, for I had none other myself'. The following were the items taken from their room, an interesting and somewhat mixed collection:

... a child's silver coral, value 8s. three yards of mode, value 18s. one man's hat, value 15s. one child's hat, value 5s. one women's gown made of silk and cotton, value 10s. three yards of linen cloth, value 4s. one yard and three quarters of cambrick, value 3s. one muslin apron, value 3s. one muslin neckcloth, value 2s. one cloth cloak, value 2s. one linen shawl, value 1s. one black silk bonnet, value 1s. two linen handkerchiefs, value 1s. one linen shift, value 1s. two womens caps, value 1s. one linen apron, value 6d. one silver watch, value 1l. 10s. and one Guinea and a half.

79. December 1792. Mrs Mary Blackmore, wealthy widow; she resided in Tabernacle Row, Moorfields. While she was at worship at the Tabernacle her house was robbed of about thirty-pounds worth of silver and gold plate and jewellery, in addition to a considerable amount of clothing:

a black silk cloak, value 2l. a pair of stays, value 1l. a silk petticoat, value 10s. a stuff petticoat, value 10s. three linen handkerchiefs, value 11s. five muslin aprons, value 2l. 5s. a linen apron, value 2s. a muslin tucker, value 6d. a muslin shawl, value 2s. a cotton shawl, value 3s. two muslin half handkerchiefs, value 6[?]. two lawn half handkerchiefs, value 6s. one half yard of thread lace, value 3s. three linen shifts, value 5s. and three guineas in monies...

80. December 1792. Mary Richardson, servant; she was a servant to 'Robert Bevis, Sloan-street, Chelsea; I lost the articles mentioned...as I was coming from Stapleton in Essex, in a single horse chaise with my master.' Mary Richardson's trunk had been tied behind the chaise and a man had run behind and cut it off while the chaise moved along Fleet Street. Mary Richardson may have been an upper servant in Robert Bevis's household.

... one wooden trunk, value 2s. one deal box, value 1l. four cheque aprons, value 6d. two linen shifts, value 7s. three cotton half shawls, value 2s. two printed cotton gowns, value 7s. one dimity petticoat, value 4s. one stuff petticoat, value 7s. one linen handkerchief, value 1s. two muslin half shawls, value 1s. one muslin apron, value 4s. four muslin caps, value 2s. one linen night cap, value 6d. one cloth apron, value 2s. one pair of lawn robbins, value 1s. one muslin tucker, value 1s. one printed muslin half shawl, value 1s. one printed bound book, value 1s. one printed half bound book, value 1s.

81. January 1793. Charles Silk, 'prepares the plated work for the coach harness'; Silk and his wife Norah lived in a house he owned at No. 7 Middlesex Court, Drury Lane. While Norah Silk was across the street visiting, their house was broken open. Norah had not locked the front door, because as she said, 'I was only gone over the way', but she had locked the parlour door, which led to the bedroom—to no avail. She estimated that five-guineas' worth of their possessions were taken, including:

... two cloth coats, value 2l. 10s. two muslin neckcloths, value 1s. a silver watch, value 1l. a watch spring, value 1d. six teaspoons, value 6d. a muslin cap, value 1d. a linen cloth cap, value 1d. a cotton gown, value 9d. a muslin shawl, value 1s. ...

82. January 1793. John White, labourer; Mary White, washerwoman; John and his wife Mary lived in Bromley, where White owned one of a row of houses. The house shared a backyard and 'other yards of other houses communicated to it, a whole row of houses communicate to it'. A large quantity of clothing was stolen, but only some of it belonged to the Whites, the rest was brought to be washed by Mary White and her servant, probably from the surrounding neighbourhood.

... seven linen shirts, value 14s. three pair of silk stockings, value 9s. four pair of cotton stockings, value 8s. three callico shirts, value 12s. four pair of misello pockets, value 3s. a flannel waistcoat, value 1s. a callico bed gown, value 2s. two dimity petticoats, value 5s. a callico bed gown, value 2s. two muslin aprons, value 4s. four callico aprons, value 4s. two check linen aprons, value 4s. one pair of pillow cases, value 3s. two diaper towels, value 2s. three pair of lace ruffles, value 6s. four muslin caps, value 4s. a muslin tucker, value 1s. a cotton night cap, value 6s[?] five linen table cloths, value 15s. a cotton gown, value 8s. two cloth coats, value 10s. a pair of cloth breeches, value 2s. a linen napkin, value 4s. a muslin handkerchief, value 6d.

Sources: * *The Public Advertiser*, 1762–6; *Proceedings of the King's Commission of the Peace ... for the City of London and ... the County of Middlesex; held at the Old Bailey*, 1732–93 (Old Bailey Records).

APPENDIX 3

Travels of Joseph Harper

	Winter/Spring Travel		Summer/Autumn Travel	
1734				
	Rugby	17–18 Jan.; 24 Feb.; 3 Mar.; 2 Apr.; 30 Apr.	Coventry Fair	13 June
	Coventry Fair	8 Mar.	Rugby	5 June; 27 June; 10 July; 19 Sept.; 29 Oct.
	Coventry	23 Apr.	Tamworth	6 June
	Oxford	25 Apr.	Lilbourne	11 July
	London	13 Mar. left; arrived 15 Mar.; stayed until 22 Mar.	Lichfield	23 Aug.
			Atherstone Fair	9 Sept.
			Hinckley Corn Fair	20–21 Oct.
			Rugby Fair	11 Nov.
1735				
	Lutterworth	30 Jan.	Atherstone	3 June; 10 June; 17 June; 24 June; 1 July; 8 July; 15 July; 22 July; 5 Aug.; 19 Aug.; 26 Aug.; 2 Sept.; 10 Sept.; 30 Sept.
	Rugby	31 Jan.		
	Kettering	1 Feb.		
	Arbury Hall	5 Feb.; 18 Feb.	Atherstone Rag Fair	10–11 Aug.
	Northampton	6–7 Feb.	Rugby	19 June; 13 Nov.

Appendix 3 *(continued)*

	Winter/Spring Travel		Summer/Autumn Travel	
	Atherstone Fair	27 Mar.	Hinckley Fair	15 Aug.
	Atherstone	13 May; 20 May; 27 May	Market Harborough	1 Sept.
	Coventry	9 Apr.	Coventry Fair	20–23 Oct.
	London	15–28 Apr.	Coventry	4 Nov.
	Nuneaton Fair	3 May		
1736				
	Arbury Hall	6 Jan.	Hinckley Fair	7 June
	Atherstone	13 Jan.; 10 Feb; 14 Feb.; 2 Mar.	Branston	8 June
	Warwick	14 Jan.; 8 Apr.	Coventry Fair	25 June
	Coventry Fair	19 Mar.	Rugby	7 July; 4 Aug.
	London	22–27 Mar.	London	7 Aug, left; arrived 9 Aug.; stayed until 16 Aug.; 25 Oct.–1 Nov.
	Coventry	4 Apr.		
	Leamington	9 Apr.		
	Tamworth Fair	23 Apr.		
	Bosworth Fair	27 Apr.		
1737				
	London	20 Jan.; 20 Feb.	Coventry Fair	10 June
	Coventry	24 Mar.	Lutterworth	25 Aug.
			Rugby	27 Aug.
			Manchester	15–19 Sept.
			Newcastle	20 Sept.
			Coventry	20 Oct.

APPENDIX 4

Customers Served by the Anonymous Manchester Firm, 1773–1779

ENGLAND

Ashbourne:
Mrs Hawkins

Bakewell:
Mrs Bullock

Bewdley:
Mrs Ann Scrimshipe

Birmingham:
Thomas Warren
Thomas Careless
William Barns
William Fitter

Blakeley, Lancs:
John Tetlow
William Dawson

Bolton:
Mrs Horridge

Bristol:
Tagart & Green
Parsons & Studley
Peter Goodwin

Bury:
George Ormerhead
George Ormrod

Buxton:
Earl of Sussex
Brian Hodgson
James Parris
John Greenhalgh
Mrs Hodgson
Revd. Robert Thorpe

James Brock
Mrs Brock
Nelly Bath
Thomas Bath
Nelly Burton
Mr Dawson
William Bott
Elizabeth Brandereth

Chadwick:
Roe & Kershaw

Chester:
William Wood
Thomas Garrat
Ralph Wilcoxson

Chesterfield:
Mrs Hodgeson
Robert Hodgeston
Mrs Swettenham

Derby:
Thomas Lowe

Eccles:
Revd. Mr Crookhall
George Ashcroft
Mrs Royle

Exeter:
John Read
Roger Rowe
Richard Davies

Halifax:
John Made

Heaton, Lancs.:
Sarah Partington

Hull:
Samuel Hall

Kings Cliffe nr. Stamford:
J. Weldon

Leake nr. Boston:
Mitchell Rasor

Leeds:
James Hazelton
Thomas & Samuel Tles[?]

Leicester:
Joseph Noble

Lincoln:
John Kent

Liverpool:
Henry Wharton
James Brownbill
John & Samuel Livesey
Messrs Lakes
Joseph Broster
Rodney Taylor
Robert & Matthew Nicholson
Captain William Ainsworth
Kenwright & Sutton

London:
H. H. Deacon
John Hankinson
Thomas Jones
Daniel Cookson
Henry Evans
John Augustus Street
John Shaw
Thomas Portens
Ellis Needham
John Shuttleworth
John Nickson (Nixon)
James Mangnall
Kettle & Mandeville
William Robinson
Lewis & Worsley
Edward Rogers
John Ticknell
Mrs Fernyhough

Yates & Miller
J. & E. Kenworthy
Marsh Reeve & Co.

Longnor, Staffs.:
Thomas Oliver

Ludlow, Salop:
Thomas Gerrard

Manchester:
Mrs Hume
William Pilling
Leigh(s) & Darwell
Edward Place
Dawson & Clegg
William Heywook (Hunts Bank)
William Wilde (Hunts Bank)
James Smith
Benjamin Bancroft
Grant & Edge
William Hanson
William Hampson
James Clough
William Arrowsmith
Samuel White & Co.
Thomas Slack
Thomas & John Tipping
John Hardman
Henry Worral
Josiah Birch & Son
Thomas Johnson (High St.)
Thomas Johnson (Market St.)
William Birch
Thomas Marriot
Mrs Hodson
John Hadfield
Robert & Nathanial Hyde
Isaac Moss
Robert Saxon
Richard Tibson
Richard Mather
Joseph Rigby
James Touchet
William Wood
Robert Callow

Messrs Bartons
James Swift
Lowe, Bate & Wright
John Heywood (King St.)
John Heywood (Market St. Lane)
Bentley & Boardman [?]
Dinwiddie, James & Gilbert
Low, Marsh & Low

Melton Mowbray:
Josiah Noble

Milnthorpe, Westmorland:
Cragg & Son

Newport Pagnell:
Josiah Bugbee
William Bugbee

Nottingham:
Taylor & Almond
Benjamin & William Waddington

Oldham:
Thomas Hobson sen.
Thomas Hobson jun.
Misses Cleggs

Openshaw:
Mrs Field

Penketh:
John Richardson

Plymouth:
Leonard Arthur

Rochdale:
James Lord
John Stitt
Richard Gore
John Kershaw (& in Stockport)

Salford:
Mrs MacNeale

Sankey:
Samuel Lomax

Sowerby:
John Tattersall

Stockport:
William Fowden
James Standering
David Hyde
—— Mayers
James Gee
John Kershaw

Wakefield:
Dr Richardson

Warrington:
Thomas Morris
Richard Clarke

Wigan:
Thomas Barton
Catherine Woods
Mrs Fogg
George Hodson
Miss Hodson

Wirksworth:
Francis Harding

Woburn, Beds.:
James Hallowill

York:
Hugh Robinson

ISLE OF MAN

Peel:
Sandford

SCOTLAND

Beeth nr. Glasgow:
Robert Stevenson

Dalkeith:
Richardson
John & William Wordlaws

Edinburgh:
James Russell

Glasgow:
James Campbell
John Burns
William & James Dowglass
Bogle & Scott

Greenock:
Robert Williamson

IRELAND

Donegal:
James Richardson

Dublin:
William Richardson

Newry:
James Lawson

Sligo:
James Winterscale

Wexford:
John Walmsley Holme

NETHERLANDS

Amsterdam:
Benjamin Bothomley (& Widow Bothomley)
Box & Co.

UNITED STATES OF AMERICA

Philadelphia:
Price & Salmon

INDEX

advertising:
  comments on perfection of advertising by Dr Johnson 60
  exemplified by Sheridan's Mr Puff 59
  expanded through medium of handbills, trade-cards, and newspapers 56–61
  George Packwood's creative advertising 59–60
  'propagandists of consumption' 60

barter, *see* second-hand trade
Behn, Aphra:
  comment on fashions 10
Boswell, James 165–6, 177
Braudel, Fernand:
  costume as language 162
Bristol 133, 196
  directory lists used-clothes dealers 74
  secondary distribution centre for Finney and Davenport 116–17, 119, 124
Buck, Anne 166

calico 4, 13, 16, 17, 19, 20
  British-made calicoes commonplace in clothing 106
  calico campaign 21, 29–31, 34–41
  calico chasing 36, 37, 39
  craze for 16, 20, 21, 33
  curiosity value 12
  imports in 1664 15
  inconsistent with the national interests 4
  production of English printed grows 32
  proposed ban 29
  worn by ladies 16
  worn in shirts and shifts in late seventeenth century 18
Cary, John 21, 27, 30
Chapman, S. D. 131
Charles II 6, 10, 12, 13, 15
  accepted payments from East India Company 15
  adopted oriental textiles 13
Chartres, J. A. 125

checks 113, 140, 190, 191
  clothing check in all cotton and cotton/linen variety 81
  'Cotton Check', seven types sold in 1730s 78
  furniture checks durable and popular 81, 82, 113
  largest group of fabrics noted by Holker 79–81
  varieties increased between 1750 and 1770 82–3
  see also cotton industry, cotton, fustian, muslin
Cherydery 84, 85
  see also Indian textiles, chintz
Child, Sir Josiah 15, 19, 180
  admitted cheap calicoes displaced some wools 27
  championed Indian textiles 14
  cultivated close ties to the court and courtiers 22–3
  described Indian textiles printed in England 31
children:
  clothing provided for poor children 107
  contributed to family economy 51–2
  Samuel Crompton, worked as child at early age 52
  William Radcliffe, worked at many jobs as a child 52
chintz 4, 13, 14, 15, 17, 31, 84, 85
  bolts of chintz shipped to London weekly 124
  largest assortment in Holker's collection 83
  see also calico, Cherydery, Indian textiles
clothing, *see* dress
clothing, ready-made, *see* ready-made clothing
Coleman, D. C. 4
Commons, House of:
  1696 bill proposed to ban oriental fabrics from England 29

Commons, House of (*cont.*):
  1700 Act prohibited printed East Indian textiles 30–1
  1720 petition from cotton spinners and weavers 76–7
  1721 Act prohibited sale and use of most East Indian textiles 41
  1734 petition from Manchester for turnpike 122
  1785 petitions submitted defending pedlars as distributors of British manufacturers 136
consumerism:
  account book reflects common pattern of consumerism 110
  active response by cotton industry to consumer 77
  affected by second-hand trade 75–6
  advertising/information effective in expanding 58–61, 167–72
  attraction of consumer goods cannot be legislated 42
  cheap accessories stimulated consumerism 87
  clothing provided for poor family 106–8
  clothing and social identification, discussion of 161–3
  consumer choice expanded through second-hand trade 177
  cottons consumed widely prior to industrialization 91
  cottons consumed with greater regularity in dress 98
  cotton textiles facilitated new consumer patterns 185, 197, 199–200
  hampered by existing productive capacity *c*.1750s 81
  Johnson Sample Book reflects pattern of consumerism 110–13, 172–3
  manifested in demand for used as well as new goods 61
  mass market and, discussion of 197–200
  material life changing 100–3, 108
  new consumer industries arising 6, 18, 19
  novelty a key factor in successful commodities 109
  patterns of consumption alerted among middle classes 108–14
  patterns of consumption altered among working classes 102, 103, 105, 107, 108
  popular fashion stimulated consumer trades 190–1, 197–200
  propagated by advertising 60
  public fascination with new commodities 61
  redistribution of desired goods through second-hand trade 65, 68
  writer opposed to 61
  see also demand
Cornwall:
  trade in cottons by Finney and Davenport 116, 117, 118–19
cotton
  advantages in the market 93
  component of dress of working classes 97–108
  denounced as luxury among lower orders 96
  Johnson Sample Book reflects changing popularity of cottons 110–13
  product for the mass market 3 n.
  qualities of the fibre infinitely variable 105
  see also calico, chintz, fustian, muslin
cotton industry, British
  all-cotton fabrics made in greater numbers 94
  benefited from ban on Indian cottons 42
  bolts of chintz shipped to London weekly 124
  British consumers look to British-made cottons 78–9
  checks made in increased variety 82–3
  coaches and carriers used to supply markets 141, 148, 152–9
  consumed widely prior to industrial expansion 91
  cottons substituted for wool and leather clothing 100–3
  criteria for demand, discussion of 43
  defended industry as national 40
  demand among working classes and middle classes 55
  demand for variety cannot be satisfied *c*.1750s 81
  demand in home market not quickly satisfied 94
  dependent on British market 43, 52, 78
  distribution patterns, discussion of 146, 152, 153, 154–9
  diversification during the 1750s and 1760s 87

dress among working classes made up of more cotton 96–108
early products 77–89
fashion in cottons key to success 189, 190
few of Holker's samples all cotton 79
founded on needs and wants in Britain 163–4, 176
furniture checks durable and popular 81, 82
fustians made up largest part of trade in Britain 87
fustians substituted for wool fabrics 112
great range of printed British chintzes 83
handkerchiefs have extensive market in England 85
handkerchiefs manufactured in Lancashire 88
health improved with cotton clothing 104
Holker reported on early manufactures 79–87
Holker acknowledged British cotton velvets superior 86
hose staple in home market 87–8
Lancashire site of fustian and linen weaving 33, 78–9
links among manufacturers, wholesalers, and retailers, discussion of 118 and n.
London's role as distribution centre for cottons 152–7, 159–60
'Manchester Ware' bought with old clothes, 'Rags' 68
Manchester as distribution centre for cotton industry 115, 120, 121, 132–5, 137, 143, 146, 150–60
manufacturers imitated Indian textiles 33
markets served by second-hand trade key to development 62, 75
Midlands stocking manufacturing 88
novelty a key factor 109, 189
Parliament ignored claims of British cotton manufacturers 41, 77–8
pedlars defended as distributors of manufacturers 136
price structure of cotton textiles 94, 95–6, 106, 107, 111, 112–13
price fell with increased production 103
printers' skills basis of popularity of printed fabrics 84
ready-made clothing trade tied to cotton industry 186–90, 194, 196–7
table of raw cotton imports 54
Touchet, Samuel testified to Parliament 82
trade extensive among manufacturers 148–50
variety preceded factories production 83
variety of textiles grew 1775–1800 94–5, 104–5, 106
velvet manufacturers guard their techniques 85–6
warp for fustians of linen or Indian cotton 80
'whole Kingdom is furnished with Commodities of this Sort' 84
Crompton, Samuel:
work performed as a child 52
curtains 18
cotton bought by Benjamin Franklin 109
East India Company increased imports 17
spreading use 82

Davenant, Charles:
advised trade with cheapest source of popular novelty 27–8
Defoe, Daniel:
denounced the spread of calicoes 16
described spread of textile printing industry 32
emphasized family benefits of salaried work for women 52
Lancashire trade noted 117
road hazards described 116, 122
scolded wool manufacturers for overproduction 34
transport methods described 116
demand:
cash-based system augmented through second-hand trade 68
cultivating popular demand 56–61
demand in home market not quickly satisfied 94
distribution system sustained demand 150
domestic demand nurtured British cotton industry 43
earnings of women and children added to domestic demand among working classes 53–5

demand (*cont.*):
  fashionable products arise from demand 163
  growing for clothing, china, and domestic accessories 56
  hidden demand of second-hand trade 61–76
  increased through advertising 56–61
  mass market and fashion, discussion of 197–200
  new product sales not sole measure of demand 61
  pawnbrokers source of cash to meet demand of less affluent 70
  printed and checked fabrics very much in demand 81–4
  seasonal rhythm 154–5
  second-hand substitutes fostered wider demand 62, 65
  trade among manufacturers to meet demand for common commodities 148–50
  two-tiered 61–2
Derbyshire:
  site of stocking manufacturing 88
Devon:
  trade in cottons by Finney and Davenport described 119
directories:
  Bailey's Midland Directory 70
  reflect distribution of pawnbrokers 70–1
  The Manchester Directory for 1773 140
  Sketchley's Bristol Directory of 1775 74
  Universal British Directory 70, 73, 75, 195
  Universal Directory for 1763 130, 182
domestic system
  children employed from early age 52
  paid employment affected market demand 53–5
  produced varied cotton, cotton/linen textiles 83
  wage labour for women and children 51–2
  see also sweated labour
Dorset:
  trade in cottons by Finney and Davenport described 119
dress:
  accessories increased in quantity and quality 87
  apron 16, 90, 91, 92, 97, 98, 106, 164, 179, 183, 195
  battante 8

blouse 179
breeches 86, 93, 100, 101, 102, 104, 105, 114, 183, 194
buckles 87, 93
cap 87, 97, 113, 179, 181, 182, 183, 186, 195
cardinal, coat, cloak 11, 16, 65, 66, 90, 92, 93, 97, 98, 100, 101, 104, 114, 179, 194, 195
collar 105
construction of women's dress 166
court dress 9
cravat, neckcloth 17, 98, 190, 194
cuffs 17
doublet 11, 16
drawers 180, 182, 194
dress of working men and artisans described 92–3, 98–108
dress of working women described 92, 97–108
footwear 86
frock 90, 93, 101, 104, 180, 182, 195
fustian used in assortment of garments 86
gowns, bedgown, mantua, morning dress, *négligé*, nightgown, nightrails, riding habit, *robe de chambre, robe de nuit, sach* 7, 8, 9, 10, 14, 16, 17, 18, 90, 91, 92, 94, 97, 98, 99, 105, 106, 108, 109, 113, 164, 166, 180, 181, 187, 188, 189, 190
handkerchief 17, 88, 92, 97, 98, 99, 100, 103, 164, 186, 190
head-dress, hood, hat, bonnet 8, 17, 25, 55, 91, 92, 93, 98, 103, 179, 186, 194, 195
hoop 86
hosiery 5, 9, 17, 87, 88, 97, 99, 104, 105, 113, 114, 164, 165, 180
jacket 99, 103, 194
jerkin 86
mitts 87
muffatees 87, 195
overall 179
periwig, peruques 93, 165
petticoat 17, 18, 92, 93, 97, 99, 164, 180, 181, 193
pocket 17, 86
restrictions in dress resisted 25
riding dress 112
shift 18, 92, 98, 180
shirt 10, 17, 18, 90, 92, 103, 104, 180, 182, 192, 193, 194

Index

shoes 103, 165
simplification of dress styles 9, 16
sleeves 17
stays 86, 91
stomacher 99, 195
suit 166
textile composition of dress changed 89
trousers 114
tucker 106
value of clothing realized in second-hand trade 69
vest 11
waistcoat 10, 90, 91, 93, 100, 101, 104, 113, 114, 181, 182, 194
waistcoat shape 105, 189, 193
see also Fashion and Appendix 2

East India Company 21, 22
  ban proposed on all oriental textiles 29
  bed clothes, bed hangings and curtains imported for labouring classes 17
  Charles II given payment by Company 15
  current fashions working to their advantage 16
  exported cotton yarn for British manufacturers 33
  founded in 1600 12
  incorporation of new company 26
  promoting Indian textiles among wealthier classes 14, 15, 16
  ready-made clothing produced for English market 180, 184, 185
  sale of East Indian imports in England opposed 16, 20, 21, 24, 26, 32
  seeking Indian goods to appeal to English tastes 13
  shipped wool cloth to India 28, 29
  Tories hostile to eastern trade/commercial interests 26
  trade defended by John Cary 27
  wool interests claim competition for same markets 22
Eden, Sir Frederick 178, 179, 188
Edwards, M. M. 129
Elizabeth I, 12
Evelyn, John 11
Eversley, D. E. C. 62

fairs, *see* trade, domestic
fashion:
  appearance key to gentility 109
  blurring social divisions 16, 185, 190;
  clothing and social identification, discussion of 161–3; servants adopt fashions of upper classes 3, 9, 96
  Boswell's plan to cultivate fashionable appearance 165–6
  cotton stockings sign of fashionability 87–8
  demand for fashion news met 168–72
  dictated by Parliamentary ban 42
  English preoccupied with fashion 165
  fads fed by quick deliveries of cottons 156–7
  fashionable clothing choice of working women 55
  fashionable touches to a butcher's attire 93
  fashion's disciples followed princesses' lead 110
  fashion not controlled by manufacturers 163
  French fashions 6
  generally uniform, becoming 64
  Indian textiles sought that met European fashions 13
  information and standardized fashions 167; fashion prints/illustrations proliferated 168–71
  Johnson Sample Book reflects use made of prints 172–3
  leather breeches made obsolete by new cottons 101–2
  London as fashion centre 131, 155
  mass market and, discussion of 197–200
  mirror of history 6
  news more readily available 167–76
  popularity of new hosiery 4
  rapid alterations in patterns of fabrics 7, 8
  ready-made clothes, and 185, 188–9, 191, 192–3, 196–7
  weavers and wool industry attempt to dictate taste 38–9
France:
  employed industrial spy, John Holker 79 and n.
  established rival cotton velvet manufacture 86
Franklin, Benjamin:
  bought new patterned printed cottons for wife 109
Fraser, W. Hamish 198, 199

Index

fustian 33, 95, 140
  breeches made in 102
  choice for riding dress 112
  corduroys: Brunswick, corded Tabby, corded thickset, Genoa, Nelson, Prince of Wales, Queen, ribdeleur, ribdurant, Royal Ribb, Wild Boar Tabby, 94, 102
  eight types sold in 1730s 78
  made up largest part of trade in Britain 87
  stuff of working men's and women's clothing 90, 145
  used instead of leather or wool for men's clothing 100–3, 114
  varieties: baragon, dimity, diaper, erminetts, everlasting, grandurelle, herringbone, hooping, jean, jeanette, nankeen, pillow, thickset, ticking 86, 101
  variety manufactured 79, 86, 101, 102
  warp of linen or Indian cotton 80

Goldsmith, Oliver 169
Greg Mill, at Styal 55

Halifax, Lord 11
Holker, John:
  established rival French cotton velvet manufacture 86
  report to French government on British cotton industry 79–87, 124
Hollands, cotton 84, 85
  see also cotton industry
Humphreys, John (weavers' leader):
  led Spitalfields weavers in anti-calico campaign 35
  pilloried 37
  sedition, accused of 35

India:
  as source of textiles 12
Indian textiles
  attacks on wearers of calico 36, 37, 39
  bed clothes and bed hangings developed as new import 17, 18
  bettilies 17
  chercanneys 18
  derriband 18
  found uses among the poor 18
  guide published in 1696 for English consumers 17, 19
  hostility of wool industry 24, 29, 39–42
  Pepys rented an Indian silk morning gown 15
  printed in England 31–2
  printed textiles banned 1700 31
  prohibition of 1700 ineffective 32
  prohibited sale and use of most cotton textiles 41–2
  renewed campaign against Indian imports 34–41
  sale opposed in England 20, 32
  Scotland and Wales supplied with Indian textiles 19
  supplanting European fabrics 14
  varieties used in late seventeenth century 18, 19
  see also calico, chintz, muslin

Johnson, Barbara 110–13, 172–3
Johnson, Dr Samuel
  comments on advertising 60

Kalm, Per 164–5
King, Gregory 43

Lancashire 126, 133, 191
  domestic trade described 117
  manufacturers produced handkerchiefs 88
  site of early linen and fustian manufacturing 33, 78–9
Lindert, Peter:
  revised wage-scale 47–9
linen 186, 190
  advantages in the market 93
  consumers depended on British linen and cotton/linen fabrics 78, 92–3
  examples in dress of working classes 97
  Holker's 1750s samples showed pure linen and mixed linen fabrics 79, 86
  Johnson Sample Book reflects competition with cottons 111, 112, 113
  Lancashire-made linen handkerchiefs 85, 88
  linen clothing a staple over most of eighteenth century 90
  linen fabrics described in 1696 guide 17
  linen warp used to make many fustians 80
  manufacturers of printed linen and cotton/linens petitioned Parliament to protect their trade 40–1
  weavers attacked wearers of printed linens as well as calicoes 36, 38–9

Index

London:
  centre of fashion, influenced trade 131, 155
  centre of silk handkerchief manufacture 85
  competing with Manchester as distribution centre for cotton industry 121, 152–7
  hub of national road network 121–2
  hub of second-hand trade 69, 72–3
  largest market for ready-made clothing 187
  middlemen in textile trade discussed 130–2, 153–5
  neighbourhoods where second-hand trade flourished 72–3
  principal wholesale centre for hosiery 116, 132
Lords, House of:
  1696 foiled proposed ban on East Indian fabrics 29
  foiled proposed ban on all printed fabrics 41
Louis XIV 6, 11, 12
luxury:
  of common people denounced 30, 96
  materialism of advertising denounced 61

McKendrick, Neil 5, 52
  earnings of wife and children added to surplus in family 53
  fashion patterns 162
  role of advertising 59–60
Manchester:
  distribution centre for cotton industry 120, 121, 126–8, 132–5
  Manchester and vicinity petitioned for turnpike 122
Manchester firm, unknown 146–59, 187–90
Mandeville, Bernard 14, 20
Mann, Julia de Lacy 31
manufacturers and middlemen
  Barton, Messrs, fustian manufacturers and supplied Dent 140, 141; traded with unknown Manchester firm 149
  in cotton trade 116–20, 123, 129–35
  dominance of Manchester-based suppliers 143–6
  Finney, Samuel and Davenport, Ralph manufacturers and merchants 116–20, 123, 124, 134
  Gough, Joseph, fustian manufacturer and supplied Dent 140
  Hanson, William, check manufacturer traded with unknown Manchester firm 149
  Hardman, John, fustian manufacturer traded with unknown Manchester firm 149
  Haworth, Jonathan & Sons supplied Thomas 144
  Hyde, Robert and Nathaniel, check manufacturers and merchants, trade of 148–9, 187
  Kenworthy, J. & E., London customer of unknown Manchester firm 153, 154
  Kettle & Mandeville, London customer of unknown Manchester firm 153, 154–5
  Lawrence, John, check manufacturer and supplied Dent 140; traded with unknown Manchester firm 149
  links among manufacturers, wholesalers, retailers discussed 118 and n.
  Livesey, Hargreaves & Company, supplied Thomas 144
  London-based middlemen in textile trade 130–2
  London's role as distribution centre for cottons 152–7
  London warehousemen listed, customers of unknown Manchester firm 153
  Manchester-based wholesalers limited London's role in cotton trade 115, 143, 144, 159–60
  Marsh, Reeve & Company, London customer of unknown Manchester firm 153, 154
  Mather, Richard, silk and fustian manufacturer and supplied Dent 140, 141
  Oldknow, Samuel manufacturer and merchant 134, 156, 189
  patterns of national distribution of cotton textiles discussed 146, 152, 153, 154–9
  Peel, Robert & Company supplied Thomas 144
  Peel, Yates, Tipping & Halliwell supplied Thomas 144
  Philips, J. & N. & Company manufacturers and merchants 132–3, 134, 149
  Place, Edward, check manufacturer traded with unknown Manchester firm 149

manufacturers and middlemen (*cont.*):
   Ridings, Samuel and John, supplied Turner and southeast 142–3
   Rideout, John and Thomas supplied Thomas 144
   Salte, Samuel linen-draper 129, 134; encouraged novelty in manufacturer 189; urged speed to meet new demand 156–7
   Totterdell, John, West-country wool-draper and clothes manufacturer 197
   Touchet, Samuel 82, 118
   unknown Manchester firm, manufacturer and merchant 146–59; ready-made clothes trade in Britain 187–90; retail customers 146–7; wholesale customers, characteristics of 147–59
   Usher, T. & Company supplied Thomas 144
   Webster, George supplied Dent 140
   White, Samuel & Company, fustian manufacturer and chapman, traded with unknown Manchester firm 149–50
   see also pedlars, shops, retailers
Mary II, Queen 25
Mayhew, Henry 67
mercantilism:
   doctrines 4
   theory undercut by Eastern trade 4
Monmouth Street, *see* second-hand trade)
muslin 4, 13, 16, 17, 19
   British-made muslin commonly used in clothing 106, 108, 110, 114
   Indian muslin continued to be bought after 1721 by the wealthy 78
   used in dress of working classes 97

new draperies 4, 6, 23, 33
   diffusion of the new product 4
   reason for their success 4
newspapers/periodicals:
   advertisers reaching all parts of nation 57–8
   agents of commercial activity 58
   Bristol Oracle 58
   Derby Mercury 58
   The English Ladies Pocket Companion 173
   Gallery of Fashion 175
   Gazetteer and New Daily Advertiser 59, 69
   General Advertiser 59
   The Gentleman's Magazine 169
   Jackson's Oxford Journal 57, 67
   The Lady's Magazine 61, 169, 170, 171, 174
   Lady's Monthly Museum 175
   Lane's Pocket Book 173
   London Magazine 96
   Liverpool Chronicle 58
   London Evening Post 58
   Magazine à la Mode 174, 175
   Mercure Galant 7
   New Royal Pocket Companion 173
   Public Advertiser 59, 172
   Punch 182
   publishers promoted the ideals of dress 168–76
   Reading Mercury 57, 58
   Read's Weekly Journal 104
   Sherbourne Mercury 171
   Stamford Mercury 68
   York Courant 58
North, Sir Dudley 20
Norwich:
   wool fabrics 79
Nottingham 126
   site of stocking manufacturing 88

Old Bailey Records
   documented clothing of cross-section of society 89, 90, 91, 113
   see also Appendix 2
Orléans, Duchesse d' 9, 11
Owen, Robert 155

pamphlets:
   advocating ban on East Indian textiles 25, 29, 38–9, 40
   disputing outcome of 1721 ban on Indian textiles 42
   supporting domestic sales of East Indian textiles 25, 27, 28, 30, 37
Pawson, Eric 157
Peach, John, Bristol merchant:
   trade ready-made attire 186
Peach & Pierce, Bristol merchants 81, 82, 83
   trade in ready-made attire 186
pedlars:
   based around cotton manufacturing towns 137
   Dent supplied by chapmen/pedlars 139–40

Index

important agents in domestic trade 135–7
Scotchmen with £1000s in trade 136
trade defended to Parliament 136
Pepys, Elizabeth 13
Pepys, Samuel:
  admired Charles II's Oriental vest 11
  fashion conscious 13, 14
  indignant on behalf of Charles II 12
  portrait painted in rented East Indian gown 15
  receives a gift of an Indian gown for his wife 14
Petticoat Lane, *see* second-hand trade
Philips, J. & N. & Company 118 and n., 132–3
Place, Francis 69
  chronicled decline of the leather breeches trade 101–2
  making ready-made breeches 183
Plumb, J. H. 163
population:
  growth as it affected real wages 45–6
  marriage patterns considered 45–6
  structure considered 43–6
  table for eighteenth century 44
pottery:
  exchanged for old clothing 67–8
printing industry, textile:
  copperplate printing noted by Benjamin Franklin 109
  defended pedlars 136
  expanded in England in 1670s 31–2
  growing range of British chintzes 83
  London printers producing most chintzes 83–4
  printers' shops attacked 36
  printers' skill basis of popularity of printed fabrics 84
  production of English printed calico by 1711 32
  skills required for textile printers 84
Proceedings of the King's Commission of the Peace, see *Old Bailey Records*
prohibition:
  of all pure cottons and most Indian textiles, 1721 41, 77, 90
  of printed Indian textiles, 1700 30–1
  repeal of prohibition, 1774 41 n.

Radcliffe, William:
  jobs performed as a child 52
Rag Fair, *see* second-hand trade

ready-made clothing:
  cotton industry and ready-made trade 185–90
  discussion of trade 178–85, 190–7
  East India Company marketed 180
  gowns bought for servants 105
  gowns sold 145
  market for ready-made clothes 191–5
  Morgan sold cotton shirts ready-made 103
  national market 195–7
  popular fashion and 185, 189, 191, 192–3, 196–7
  popular item distributed by unknown Manchester firm 154–5, 156–7
  produced items of domestic manufacture 180
  sold by pedlars and in shops 180–2
  slops and clothing for working men 184–5
  sweated labour component of trade 182–3
  variety sold by unknown Manchester firm 187–90
Rosemary Lane, *see* second-hand trade

salesman, *see* second-hand trade
Schofield, R. S.:
  population structure 43–6
second-hand trade 61–76, 176–9
  barter common strategy 62, 64, 67
  Bristol traders listed in directory 74
  cash for customers 70
  characteristics 61–4
  discussion of, as affected demand 61–3
  Exchange established 1843 73
  expanded consumer choice 177
  London hub of second-hand trade 69, 72–4
  John Matthews, salesman with national trade 67
  Monmouth Street, site of second-hand trade 64
  new goods bought with old 67–8, 69, 70
  organized system of redistribution 62
  pawnbrokers with local and national trade in used clothing 70–2
  Petticoat Lane, site of second-hand trade 64
  pottery exchanged for old clothing 67–8
  provincial dealers supplied from London 73–4

second-hand trade (*cont.*):
  Rag Fair, site of second-hand trade 71
  redistribution network for clothes 65
  reflected wider level of demand 62
  Rosemary Lane, site of second-hand trade 64
  salesmen/women, practitioners in trade 63 n.
  served all Britain, colonies, and Europe 65, 67, 74–5
  shopkeepers recycled textiles and clothing 65–6
  Strype, John described development of trade in London 72–3
  substitution of used goods meets wide demand 62–3
  Susannah Somers of Biggleswade, Beds., saleswoman 74–5
  tailors accepted trade-in of old clothes for new 69
  trade commonplace 62–3
  women clothes dealers 71–2
servants:
  attempt to legislate clothing 25
  best dressed of wage-earners 96
  chambermaids claimed to dress like mistresses 16, 190
  dress of female servant described 97
  ready-made gowns bought for 105
  substituting Indian fabrics for wools 38–9
shops, retailers
  Allin, John, Birmingham tailor and salesman 196, 199
  Atkinson, Elizabeth, Manchester milliner and linen-draper 147–8
  Barker, Harry, London linen-draper 191; supplier of ready-made clothing
  Blunt, Robert, London shirt warehouseman 192, 193
  Bothomley, Benjamin, Amsterdam customer of unknown Machester firm 151
  Bromley, London linen and shirt warehouse 193
  Brown, Elizabeth, Norwich milliner 195
  Bryan, Mr, London tailor and dealer in ready-made 192
  Bryant, Mr, London coat and cloak warehouseman 194
  Dent, Abraham, Kirkby Stephen shopkeeper 65, 101, 138–42, 144; supplied with cottons by Manchester-based firms 139–41; supplied by chapmen/pedlars 139–40
  Dowglass, William and John, Glasgow customers of unknown Manchester firm 151–2
  Harper, Joseph, Hinckley shopkeeper and tradesman 120–1
  Hogarth, Mary and Ann, London linen-drapers 181–2
  Forster, London linen warehouseman and dealer in ready-made 192
  Lamonby, Joseph, Wirksworth clothes shopkeeper 196
  London shops carried variety of cotton shirts 103
  Morgan, John and Mary, Welsh drapers 102, 103, 104, 113, 145–6, 196–7; carried Manchester stock 145–6
  recycling textiles and clothing 66
  Roberts, Mrs, London dealer in ready-made clothes 192
  selling British cottons 101, 102–3
  ready-made clothing sold 181–2, 190–7
  supplied most clothing 179, 183–4
  Taylor, Mr, London quilted coat warehouseman 193
  Taylor & Almond, Nottingham customers of unknown Manchester firm 150–1
  Thomas, John, Hinckley dry-goods and drapery retailer 142–4; supplied with cottons by Manchester-based firms 143–4
  Turner, Thomas, East Hoathley shopkeeper 66, 142–3, 144
  see also middlemen; pedlars; trade, domestic
silk 4, 6, 14, 108, 157, 188
  French 7, 13
  handkerchiefs made in London 85
  Johnson Sample Book showed silk predominant, 1753–79 111
  Pepys rented a silk morning gown 15
  silk supplanted by Indian fabrics 16
  silk patterns and pace of development 7
  silk and wool products 79
  Spitalfields weavers 35

Index

Spitalfields 35, 79
Somerset:
   trade in cottons by Finney and Davenport described 119
Spitalfields, *see* silk, weavers, wool industry
Spufford, Margaret 65, 180, 181
Stow, J. 5
Strype, John 72
Stuart, James (Pretender) 35
sumptuary legislation 6, 9
   1721 ban on Indian textiles seen as species of 42
sweated labour:
   feature of ready-made clothing trade 182–3

tailors:
   accepted trade-in of old clothes for new 69
textiles, *see* calico, cotton, cotton industry fustian, Indian textiles
Thirsk, Joan 8, 18, 61, 176, 180, 198
trade, domestic
   carrier services, role of 123–6, 128, 141, 152–5, 157–9
   coastal shipping dependent on weather 123
   distribution patterns of cotton textiles discussed 145–6, 152, 153, 154–9
   fairs as part of trade cycle 120
   improved highways lead to improved trade 123
   London-based distributors faced successful competition from Manchester distributors 143, 144, 159–60
   London's role a distribution centre for cottons 152–7
   Manchester developed as distribution centre for cotton industry 120, 121, 132–5, 137, 143, 146, 150–60
   Manchester Men, role in trade 132
   pattern of ordering varied with distance from supplier 150–2
   pedlars important component of domestic trade 135–7
   roads as trade routes and hazards 115–16, 122
   road carriage preferred by merchants for reliability 122–3
   trade extensive among manufacturers 148–50
   transport methods described 116
   warehouse facilities, development of 124–6
   see also middlemen, pedlars, shops, retailers
trade, foreign:
   Beekman, James supplied by Bristol merchants Peach & Pierce 81, 82–3
   Beekman, James supplied by Manchester manufacturers and merchants 148–9
   claimed East Indian trade could lead to invention 30
   critique of foreign trade which hampers English exports 20
   unknown Manchester firm's foreign trade 148
transportation:
   carrier and coach service develop 121–3
   carrier service and coastal ships used by Finney and Davenport 119
   coach haulage and pattern of use 155–7
   costs of transportation, discussion of 157–9
   improvements in 128
   methods of transport for domestic trade 116
   Pickford's carrier service development as adjunct to cotton industry 123–4, 126, 152
   roads as obstacles to trade 122
   road-based distribution infrastructure established 124–8
   road travel preferred by domestic traders 122
   social change wrought by improved transportation 128
   travel times decreased 156
Tristan, Flora 162, 163
Turnbull, G. L. 123

Unwin, George 129

velvet 6
   all-cotton the largest category cotton textile 85–6
   jealously guarded techniques among manufacturers 85–6
   supplied to shopkeeper 140
   velvets, velverets, and velveteens popular with consumers 100, 101, 102, 113, 114

Wadsworth, A. P. 31

wages:
  as affected by population growth 45–6
  Britain a patch-work of wage rates 48
  declining real wages countered with longer hours of work 51
  fall averted through industrial growth 46
  family income rises with paid work for women and children 51–2
  influences on wage rates 47–51
  money wages comprised only part of payments 50
  payment in kind 50
  wage-earners also consumers 52–3
War of Spanish Succession (1701–13): disrupted wool exports 34
Weatherill, Lorna 56, 82
weavers:
  attempt to enforce public dress 38
  chased wearers of Indian calico 35, 36, 37
  Company of Weavers of London appealed for calm 37
  riot and disturbance 35–7
  silk and wool weavers besieged Parliament 29, 35
  Norwich weavers joined protest 36
  seditious weavers 37
  Spitalfields weavers riot against calico 35
  see also John Humphreys, weavers' leader
Wilkes, John:
  paraphernalia manufactured during Wilkes campaign 88
Willan, T. S. 65, 137, 138
William III 24
Williamson, Jeffrey:
  revised wage-scale 47–9
women:
  Atkinson, Elizabeth, milliner and linen-draper 147–8
  Baxter, Mary, carrier 126
  Brown, Elizabeth, milliner 195
  dress of working women described for mid-century 92
  figured as consumers 53, 54, 55
  Hogarth, Mary and Ann, linen-drapers 181–2
  Johnson Sample Book reflects preoccupation with fashion 110–13, 172–3
  joined in wage economy 51
  Mrs Moses, used clothes dealer 71–2
  Mrs Nowler, pawnbroker 72
  salaried women improve standard of living in family 52
  Susannah Somers of Biggleswade, Beds., saleswoman 74–5
  varied employment 54
Woodforde, Revd James 177
  bought from pedlars 135–6
  chose cotton for his niece 109
wool industry:
  Act requiring burial in wool from 1678 24
  campaign against Indian imports 34–41
  Cockayne débâcle 23
  competing with European manufacturers for markets 23
  Charles Davenant critical of wool interests 27, 28
  decay of trade claimed 24, 29, 39
  English calico imitations competing with wool 33
  excused weavers' riotous behaviour 37
  foresaw resurrection of wool after 1721 ban 41–2
  health of wool next the skin challenged 104
  hostile to East Indian textiles 24
  Johnson Sample Book shows reduced use of wool 111, 112
  legislation to boost domestic consumption of wool fabrics 25
  opposed to all printed fabrics worn in England 38
  prohibition of 1700 ineffective 32–3
  roughness and poor colours of wool disputed 21
  substituted for cotton textiles 100
  war disrupted exports 34
  worn by poor 107–8
  wool supplanted by Indian fabrics 16
  worsted stockings seen only on poorest working classes 104
  see also new draperies, weavers
Wrigley, E. A.:
  population structure 43–6

Yorkshire 126
  light-weight wool products 79